D1222837

Q
158.5
.C66
1979

EDITORIAL RESEARCH REPORTS ON

Advances

IN

Science

WITHDRAWN

Timely Reports to Keep
Journalists, Scholars and the Public
Abreast of Developing Issues, Events and Trends

Published by Congressional Quarterly Inc.
1414 22nd Street N.W.
Washington, D.C. 20037

Tennessee Tech. Library
Cookeville, Tenn.

292607

About the Cover

The cover was designed by Richard Pottern, art director of Editorial Research Reports.

Copyright 1979 by Congressional Quarterly Inc.
published by Editorial Research Reports

The right to make direct commercial use of material in Editorial Research Reports is strictly reserved for magazine, newspaper, radio and television clients of the service.

PRINTED IN THE UNITED STATES OF AMERICA, FEBRUARY 1979

Editor, Hoyt Gimlin
Associate Editor, Sandra Stencel
Editorial Assistants, Patricia Ochs, Diane Huffman
Production Manager, I.D. Fuller
Assistant Production Manager, Maceo Mayo

Library of Congress Cataloging in Publication Data

Congressional Quarterly, Inc.
Editorial research reports on advances in science.

Bibliography.
Includes index.
1. Science.   I. Title.   II. Title: Advances in science.
Q158.5.C66 1979   500   78-25601
ISBN 0-87187-142-4

Contents

Contents

Foreword

The Louis Harris polling organization last year asked Americans what made their country great in the past. Of those responding to the question, the overwhelming majority — 89 percent — put scientific research at the top of the list. An even greater number — 91 percent — said that scientific research would be a major factor in America's future success.

Despite this strong show of support, many scientists believe they are in trouble with the public. What these scientists find most disturbing is the growing demand for public participation in scientific decision making. They fear this trend could lead to less freedom for them, less money for their work and fewer benefits for society.

The idea that the public should somehow monitor and possibly curb scientific research grew out of a variety of sources, including the activist attitudes of the 1960s, the deepening mistrust of all established institutions and the growing realization that the application of new knowledge is not always to the good. In this book's opening report, Kennedy Maize links the demands for greater public control to the "vast public funding for science." Right now the federal government supports about 60 percent of the nation's basic research effort.

The issue of public participation in science will be subject to debate for years to come. Meanwhile, public influences already have affected the direction of scientific inquiry. Under mounting pressure to match the dramatic achievements in other areas of science and medicine, legislators have poured billions of dollars into the still-futile effort to defeat cancer. Public interest and support also have influenced the development of pollution control technologies and alternative energy sources. For most Americans the ultimate test for science appears to be the extent to which it contributes to a better quality of life.

Sandra Stencel
Associate Editor

February 1979
Washington, D.C.

POLITICS OF SCIENCE

by

Kennedy P. Maize

**May 26
1 9 7 8**

POLITICS OF SCIENCE

MANY AMERICAN scientists believe they are in trouble with the public. In particular, they are troubled about public support for basic research — the search for new knowledge and understanding of fundamental natural phenomena and processes. Scientists sense that what was once a warm and trusting acceptance of the role of science has turned into cold skepticism. Scientists say they fear that the terms of the implicit agreement between the sciences and the public that permitted unfettered freedom and generous helpings of public funds may be changing toward greater participation of non-scientists in making science policy. And they fear that new arrangements may result in less freedom for them, less money for their work, and fewer benefits for society.

Biologist David Baltimore, a Nobel Prize laureate,[1] recently spoke of a developing backlash "in which various groups in our society question whether the freedom that has characteristically been granted to research biologists by a permissive public required modification."[2]

The National Science Board, a group of scientists appointed by the President to advise the National Science Foundation, polled the leaders of American science for a 1976 report entitled "Science at the Bicentennial: A Report from the Research Community." The scientists who responded said they believed that public attitudes toward science were increasingly hostile and could lead to conditions the scientists deplored: fewer dollars for basic research, more government red tape, and limits on the freedom of scientists to study whatever they please. Oregon State University President Robert MacVicar told the board: "I do not think that it is enough to chalk this up as some kind of temporary aberration of anti-intellectualism,...it appears to be...a very serious breach of confidence between those who must support basic science in the United States and the scientific community."[3]

So far little hard data support the scientists' fears. A poll con-

---

[1] He was awarded the Nobel Price in physiology and medicine, in 1975, for discoveries about the transfer of genetic information in cells.

[2] David Baltimore, "Limiting Science: A Biologist's Perspective," *Daedalus* (journal of the American Academy of Arts and Sciences), spring 1978, p. 37.

[3] National Science Board, "Science at the Bicentennial: A Report from the Research Community," 1976, p. 76.

ducted for the National Science Board by Opinion Research Corporation in 1977 indicated that the American public accords only physicians more prestige than scientists. Drawing on earlier polls conducted for the board, the pollsters concluded that there was "a general decline in the public's overall regard for the 10 occupations [listed in the poll] from the 1960s to the 1970s and again from 1974 to 1976." But "the relative standing of scientists has improved since the 1960s."[4] Other polls have shown similar results.

Philip Handler, president of the National Academy of Sciences, told Editorial Research Reports he believed the polls were accurate. "The data are remarkably consistent," he said. "They show that by and large the general public regards scientists as a class with high regard and confidence and feels that overall society has received great benefits from science. Science's antagonists are small groups of people who have special interests outside science — opponents of nuclear power, for example — and whose complaints are not really against science. They are really seeking changes in our society and our culture. But they find science a convenient target."

But fears persist among many scientists, reflecting an intuitive assessment of events over several years. These scientists say they find evidence of an anti-science bias in public controversies over scientific issues, in the pattern of research funding, and in the increased scrutiny that Congress gives scientific issues.

Concern Over Regulating DNA Research

Many scientists see an ominous threat to scientific inquiry in the current controversy over recombinant DNA (deoxyriboneucleic acid). This controversy arose when scientists themselves became concerned over the application of the techniques of gene splicing for the creation of new or drastically altered forms of life. Initially the DNA researchers, by common consent, halted their work while the National Institutes of Health drafted guidelines for doing their work in projects receiving federal support.[5]

But in raising the issue of the risks of recombinant DNA technology, scientists opened the door to public regulation. Congress promptly stepped through that door with attempts in the summer and fall of 1977 to enact legislation that would apply the NIH safety rules to private industry, set up a government body including non-scientists to oversee the work, and determine

[4] National Science Board, "Science Indicators 1976," 1977, p. 168. The 1977 poll listed 10 occupations or professions. In terms of prestige, the rankings were in descending order: physicians, scientists, engineers, the clergy, architects, bankers, lawyers, accountants, businessmen, members of Congress.

[5] For background on the recombinant DNA controversy, see "Genetic Research," *E.R.R.*, 1977 Vol. I, pp. 223-244.

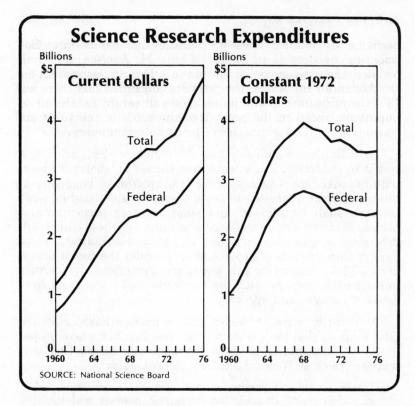

Science Research Expenditures

Billions

Current dollars

$5

4

Total

3

Federal

2

1

0

1960   64   68   72   76

Billions

Constant 1972 dollars

$5

4

Total

3

Federal

2

1

0

1960   64   68   72   76

SOURCE: National Science Board

the role of local governments in regulating the research in their jurisdictions. Sen. Edward M. Kennedy (D Mass.) introduced a restrictive bill in the Senate and Rep. Paul G. Rogers (D Fla.) offered a similar but somewhat less restrictive bill in the House.

A group of scientists, led by Harlyn O. Halverson of the American Society of Microbiology, rapidly organized to lobby against the Senate bill. The pro-DNA scientists argued that the risks had turned out to be vastly exaggerated and did not justify regulation. For example, biochemist James Watson, co-discoverer of the structure of DNA, said at a meeting NIH called to review safety procedures: "There are lots of things that scare ...me, like Tris, but recombinant DNA, no.... The dangers of this thing are so slight — you might as well worry about being licked by a dog."[6]

In the meanwhile, opponents of unrestricted recombinant DNA research were lobbying against the House bill. So far, the lobbying campaigns have produced a stalemate. Neither bill has come to a vote in Congress, and passage is not considered likely in the near future. But the fight over regulation appears to have

---

[6] Quoted by Nicholas Wade in "Gene Splicing Rules: Another Round of Debate," *Science*, Jan. 6, 1978, p. 33. Watson and co-discoverer Francis Crick, a Briton, won the 1962 Nobel Prize in physiology and medicine. Watson's account of their work appears in his book *The Double Helix* (1968).

permanently kindled fears in the scientific community that scientific freedom is in danger. Philip H. Abelson, editor of *Science,* the prestigious journal of the American Association for the Advancement of Science (AAAS), said in January: "During 1977 the scientific community escaped a threat to the freedom of inquiry in the form of harsh legislation.... The escape from restrictive legislation may prove to be only temporary."[7]

David Baltimore similarly saw the regulatory legislation as a threat to the freedom of scientists to engage in whatever line of inquiry strikes them as interesting or worthwhile. "The [Senate] bill was a clear invitation to begin the process of deciding what research shall be allowed and what research prevented.... I believe that the long-term success or failure of these efforts will determine whether America continues to have a tradition of free inquiry into matters of science or falls under the fist of orthodoxy." Those supporting this line of argument generally identify free scientific inquiry with the constitutionally protected freedoms of thought and speech.

Their critics argue, however, that scientists have created a false issue — that the question is not freedom but where proper limits should be placed to best serve the public. Sissela Bok, who teaches ethics at Harvard Medical School, said:

> These solitary, reflective, almost passive connotations of "scientific inquiry" do not in fact correspond, however, with many of the activities of today's scientists. These men and women are far from solitary in their interaction with others and in their use of vast public funds, far from passive in their use of powerful tools and machines and in the effect of their work on human welfare. They act upon nature and upon human beings in ways hitherto inconceivable. The freedom they ask to pursue their activities is, then, freedom of action, not merely freedom of thought and speech.[8]

Long-Term Decline in Research Funding

The debate over regulating recombinant DNA has occurred in the context of a general decline in government support of basic scientific research since the Vietnam War years. The growing war costs forced Congress and the Johnson administration to economize at home. Nevertheless, the decline convinced many scientists that the government did not understand the value of basic research. Their anxiety was made more acute by a rising demand in government that research be oriented more toward practical results.

Since World War II, the federal government has assumed the major role in supplying money for basic research, to the point

---

[7] Philip H. Abelson, "Recombinant DNA Legislation," *Science,* Jan. 13, 1978, p. 135.
[8] Sissela Bok, "Freedom and Risk," *Daedalus,* spring 1978, p. 116.

that currently the government in Washington supports about 60 per cent of the nation's basic research effort — $3 billion in fiscal year 1977 out of a total of $5 billion spent on basic research. Universities and colleges, foundations, and some companies provide the rest of the support. In the case of colleges, universities and foundations, their funds are limited and research has become more expensive over the years. In the case of private industry, research results often do not lead to patentable and marketable processes and products, so there is little profit in basic research.

"The freedom they [the scientists] ask to pursue their activities is...freedom of action, not merely freedom of thought and speech."

Sissela Bok
Harvard Medical School

Federal expenditures for basic research have climbed from about $500 million in 1960 to over $3 billion in 1977. But calculated in terms of the dollar's true value — after allowing for inflation — the story has been different. Judged on this basis, funding increases slowed somewhat in 1967 and began declining in 1968. The decline continued throughout the Nixon presidency but showed some improvement under President Ford. This slight upturn was carried forth by the Carter administration, both in the budget for this fiscal year and next. The 1979 budget calls for an 11 per cent increase for basic research.[9] Inflation, though rising, is expected to fall short of that figure.

Some analysts say this increase, though welcome, is not enough. Derek de Solla Price, a Yale University science historian, argued the case for science at the AAAS annual meeting in Washington in February. At that time Price told a joint meeting of the Senate Commerce Committee and House Science and Technology Committee that the long-term funding decline is "underinvestment in the future." He predicted that it would lead to a "loss in the U.S. empire in science and technology.... Academic research in science and technology has been running effectively at half speed compared with the world growth rate of 6 percent per annum increases in scientific and technological activity."[10]

---

[9] U.S. Office of Management and Budget, "The Budget for Fiscal Year 1979, Special Analysis P," pp. 309-310.
[10] Congressional testimony delivered Feb. 14, 1978. The text appears in *Science*, March 17, 1978, p. 1188.

Philip Handler has said that to support basic science at the 1967 level requires three times more funds than the Carter administration has proposed. He said that equipment is much more costly today, there are many more scientists to support than in 1967, and the research itself is more difficult. "Most of the easy questions have been dealt with," he said. But Handler acknowledged that figures much above those the administration proposes are politically unrealistic.

Sen. Proxmire's Golden Fleece Awards

The chief congressional nemesis of the scientific community is Sen. William Proxmire (D Wis.). Proxmire, head of the Independent Agencies Subcommittee of the Senate Appropriations Committee, frequently accuses the National Science Foundation of financing frivolous research projects. Proxmire also makes a monthly "Golden Fleece Award" to "the biggest or most ridiculous or most ironic example of government waste" in each 30-day period. Research projects are frequent award winners, particularly projects in the social sciences, and more particularly projects dealing with sex research. A recent Golden Fleece recipient, for example, was a project funded by the National Institute of Mental Health that included a study of the social role of a brothel in Peru.

Proxmire consistently seeks to tie research funding to specific benefits from the projects. His questions typically follow this line: "Why should the taxpayer condone this use of his money?" and "What are the long-range implications of a study such as this for a farmer in Wisconsin who is making $10,000 a year?" Both were asked at a committee hearing Feb. 24, 1977. Dr. Richard C. Atkinson, director of the National Science Foundation, answered one of Proxmire's questions with a defense of basic research:

> When one examines these trace studies and tries to determine the background information that led to a particular finding [Atkinson testified], one discovers that about 60 to 70 per cent of the key events that led to that practical development were derived from basic research results where the principal investigator had no initial conception of how the information might be applied.... The importance of an unrestrained, free choice of research projects and their direct ties to subsequent developments, sometimes two years down the road, sometimes 50 years, is well documented.

Some observers say Proxmire's approach is more concerned with the politics of re-election than it is with the nation's need for basic research. Journalist Vic Gold, for instance, said, "The deeper political thrust of the Golden Fleece proceeds from what Proxmire cynically understands as a Yahoo disdain for any long-range, egghead, scientific project that costs more than a carton of

Bull Durham roll-your-own and does not promise instant, tangible benefits."[11]

Yet Proxmire clearly operates with presumptions about science that are shared by many persons both outside and inside the scientific community. There is a feeling that science has not come up with all those "subsequent developments" Atkinson noted. Some of the developments that science heralded a few years ago, such as DDT and swine flu vaccinations,[12] have turned sour. Scientific breakthroughs in basic knowledge have resulted in technology of unprecedented destruction — such as the atom bomb. And promised breakthroughs — the well-funded conquest of cancer — have turned out, after the application of hundreds of millions of the public's dollars, to be unattainable.

Some argue that scientists themselves have contributed to the problem, by making extravagant claims for their work and by using their scientific talents in pursuit of political ends. Harvard political scientist Don K. Price focused on this aspect recently: "Politicians have been persuaded by scientists that political issues can be solved by scientific methods, and hence use political power to increase the proportion of research work that is devoted to applied problems, to increase the amount of support that is given to institutions and programs for political reasons, and to drag (or welcome) scientists into political activity. All this may work against the best interest of basic research...."[13]

The Public's Role in Science

THE RENAISSANCE marks both the birth of modern science and the origin of disputes between science and established authority. In most of 17th century Europe, authority meant the Roman Catholic Church, but as time passed the focus of the challenges to science shifted from Church to state.

Leonardo da Vinci (1452-1519), the quintessential Renaissance man — scholar, artist, engineer and inventor — was also the first modern scientist. While he contributed no important scientific discoveries of his own to the sum of knowledge, Leonardo gave Western man a way of looking and thinking that was essential to the development of science. Leonardo abandoned the traditional, Aristotelian method of using pure logic to discern the

---

[11] Vic Gold, "Calling to the Yahoo," *Harper's*, October 1976, p. 34.
[12] See "Influenza Control," *E.R.R.*, 1976 Vol. II, pp. 701-720.
[13] Don K. Price, "Endless Frontier or Bureaucratic Morass?" *Daedalus*, spring 1978, p. 86.

meaning of things; he substituted a Platonic penchant for observation.

Leonardo had a passion for detail. He haunted the hospitals of Milan so that he could study and autopsy corpses and better understand the structure of the human body. In *The Western Intellectual Tradition,* Jacob Bronowski and Bruce Mazlich said of Leonardo: "When almost all thinking was guided by universal and *a priori* plans of nature, he made a single profound discovery. He discovered that nature speaks to us in detail, and that only through the detail can we find her grand design. This is the discovery at the base of modern science, all the way from atomic structure to genetics."[14]

The first great confrontations between science and the Church took shape after Leonardo's lifetime, and centered on the work of three other Europeans — Nicholas Copernicus (1473-1543), Joannes Kepler (1571-1630) and Galileo Galilei (1564-1642). Copernicus, a Polish astronomer, offered a new theoretical explanation of the solar system that countered the traditional view based on the work of Claudius Ptolemy, the 2nd century Greco-Egyptian. The Ptolemaic picture of the universe was of a series of spheres that revolved around the earth.

Copernicus' theory that the earth revolved around the sun was revolutionary not only in conception but in its challenge to orthodox theology. Copernicus delayed publishing his work until the year he died, for fear it would outrage the religious hierarchy. But early in the following century his theory would be proven by Kepler, a German astronomer. Out of Kepler's observations he formulated three laws to describe the motions of the planets, which he published in 1609. The laws provided exact mathematical confirmation of the Copernican theory.

Galileo, an Italian like Leonardo, brought to his studies a mind thoroughly adapted to the observation of detail that Leonardo had established. Galileo's first great discovery — the use of the pendulum to measure small increments of time — marked his empirical and practical bent. Using the timing of the pendulum, Galileo was able to formulate a precise law for the fall of objects: the total distance the object travels from rest is proportional to the square of the time it has been traveling. The practical and empirical Galileo took a technology developed in Holland — lens optics — and produced a telescope to examine the heavens. He made many startling discoveries including craters on the moon, sun spots and Saturn's rings.

By the time of Galileo's discoveries, the Church was growing

---

[14] J. Bronowski and Bruce Mazlich, *The Western Intellectual Tradition* (1960), p. 18.

uneasy at the advances of science. Galileo came to symbolize the threat of the new knowledge, as he publicly began teaching the work of Copernicus. In 1616 the Church branded the theory of Copernicus a heresy and Galileo was told to cease defending it. After the publication in 1632 of Galileo's "Dialog on the Two Principal World Systems," which ridiculed defenders of the Ptolemaic system, the Italian scientist was forced to recant his "errors." That act broke Galileo's spirit, and he died in 1642, blind and under house arrest.

Scientific Inquiry by British Societies

By one of those odd quirks of history, Isaac Newton (1642-1727) was born the year Galileo died. That coincidence nicely symbolizes the movement of the center of scientific thought from Italy to England that occurred in the late 17th and early 18th centuries. The pivotal figure in the emergence of English science was Newton, and his scientific dominance came about within the context of a unique institution, the British Royal Society.

The Royal Society was chartered by King Charles II in 1662 as "The Royal Society of London for the Promotion of Natural Knowledge." The charter, though it brought no financial support with it, marked the acknowledgement by the Crown that science was of value to the state. Similarly in France in 1666 King Louis XIV chartered the Académie Royale des Sciences with much the same mission as the Royal Society.

These societies had a strong practical bias from the start. Their founders and Fellows were convinced that science was a useful enterprise. To this end, the Royal Society focused much energy on publishing the scientific works of its members so that others could share them. The Royal Society was thus interested in the language of science and stressed clear expression and simple language, prompting Bronowski and Mazlich to observe: "Their style has remained the aim of science ever since, and has proved as elusive an aim to their successors as it was to them."

Newton dominated the Royal Society during its most creative period, serving as its president from 1703 to his death in 1727. His leadership accomplished much good, and perhaps an equal amount of ill, for the society and for science. Newton's gifts were mathematical and procedural. He was able to find precise mathematical formulae to describe the nature of his observations. He was also able to devise a procedure for continually testing and validating his hypotheses, a procedure now recognized as the scientific method, the crux of modern science. But Newton also had a disastrous influence. He was not interested in the practical results of his work and looked down on those of scientific bent who were. By the time Newton died, inventors were no longer regarded as reputable scientists, and a schism between science and technology had opened.

Partly as a result of the narrow arrogance of Newton and the Royal Society, alternative scientific organizations sprang up in England. A typical example was the Lunar Society, a group of practical inventors, scientists and invention-minded industrialists from Birmingham who, between 1770 and 1800, met to talk about science. Because roads were bad and travel difficult, the society met once a month on a night near the full moon — and thus the name. Among the leading figures in the Lunar Society were Matthew Boulton, James Watt's partner in the steam engine business; Josiah Wedgwood, founder of Wedgwood potteries; and Erasmus Darwin, a doctor and biologist, the grandfather of Charles Darwin.

The Lunar Society was just one of many similar groups in England and Scotland. All were run by practical men, dedicated to improving the lot of mankind through discovery. One of the most famous members of the Lunar Society was Joseph Priestley, the chemist who first isolated and identified oxygen. It's a mark of his practical nature that Priestley immediately went on to invent the oxygen tent, used in medicine ever since.

American Patterns of Science Support

Priestley provided a transition from England to the developing field of science in the United States. Priestley left Birmingham after a rioting crowd — which believed he supported the French Revolution — destroyed his books and apparatus. In 1794 he came to America, where he died 10 years later. Science had achieved considerable respect and support in the new nation. Philadelphia was the center of American science, largely because Benjamin Franklin's Philosophical Society, modeled after English scientific societies, was located there. Also in Philadelphia were Benjamin Rush and Benjamin Silliman, both respected and important chemists and physicians. Science was also flourishing in Boston, where the American Academy of Arts and Sciences was established in 1780.

The primary goal of American science from its earliest days was practical invention, not simply the quest of abstract knowledge or truth that Newton sought. Throughout the 19th and early 20th centuries, most research was supported by private funds, and many discoveries came about in industrial laboratories. The objective of research was a patentable invention that could lead to profit. According to historian Daniel Boorstin, "In the United States the modern industrial research laboratory would eventually find the habitat where it could flourish without precedent."[15]

---

[15] Daniel J. Boorstin, *The Americans: The Democratic Experience* (1973), p. 538.

Patent Policy: Government vs. Science

If the government pays a scientist to come up with a discovery that is commercially valuable, who should own the patent rights? That question currently has Congress, the universities, and the scientific community in an intense dispute. The stakes are high, and the dispute is heated.

William Carey, executive director of the AAAS, has called the proposition that the government should own the rights "absurd," and describes it as "barring the door so the inventions can't get out." But Sen. Gaylord Nelson (D Wis.) has said that permitting the inventor or the institution where he works to patent the discovery deals the taxpayers "a one-two punch." "First they are forced to pay through the nose for this risk-free, tax-supported research and development. Then they pay dearly all over again for the grossly inflated prices these companies charge for the products they market under the patent rights given them by the government." The Carter administration has not taken a stand in the dispute.

While the federal government supports nearly two-thirds of all research in the nation, it has not established a uniform patent policy for the inventions that flow from the research. Some 22 different patent arrangements are in effect, varying widely from agency to agency. Congress has had legislation before it for over a year that would establish a federal policy. But no hearings have been held, largely because there is no agreement on how the issue should be handled.

The patent was the chief mechanism for government support of science throughout most of American history. The patent is simply a government grant of a monopoly to an invention, so that the inventor derives its economic benefits. It is no coincidence that a grandiose example of 19th century architecture was the U.S. Patent Office in Washington, D.C. It has since become the National Portrait Gallery. In American history, there are few references to 19th century researchers, but many to inventors such as Joseph McCormick and his reaper, Elias Howe and his sewing machine, and Thomas Edison and his many inventions, including the phonograph, the motion picture and the light bulb.

The government did support American science in more direct ways, although the amount of the support was quite small. The first example of this support was the establishment in 1846 of the Smithsonian Institution, organized in Washington for the promotion of science on a national scale, and financed by a gift of $500,000 from an Englishman, James Smithson. The Morrill Act of 1862 and of 1890 established ways to support American colleges and universities, especially their teaching of "the agricultural and mechanical arts." In 1863 Congress chartered the National Academy of Science to provide independent scientific advice on matters of national importance. The great landgrant universities, formed largely between 1870 and 1900, became the centers of research, fostering a science that historian A. Hunter Dupree called "independent of the vulgar uses which filled the minds of the government or industrial scientist."[16]

In the 18th century, there had been free movement between science and the liberal arts, and specialty training in the sciences had not yet evolved. Benjamin Franklin and Thomas Jefferson both combined interests in science with philosophical gifts. Priestley was a Unitarian minister and author of a pamphlet on government that influenced Jeremy Bentham. But in the 19th century, specialization became the rule, driving a wedge not only between science and the arts but between science and technology. Dupree observed: "So far apart did these two groups drift that they lost touch with one another. The basic scientist reserved his feeling of fellowship for other basic scientists like himself."

As the 20th century progressed, those divisions between science and technology decreased somewhat under the pressure of military research. By the end of World War II the schism had generally disappeared in the physical sciences, but remained a potent factor in the biological sciences. The war had meanwhile fostered a new, widespread commitment on the part of government to science.

Vannevar Bush's Plan for Postwar Era

The wartime Manhattan Project, which organized both science and technology under government control in support of the development of the atomic bomb, provided a model for the role of science and state in the postwar period. And in the aftermath of guilt over creating the bomb, American scientists took their first tentative steps into the politics of dissent.[17]

The architect of the partnership between science and government that has come to typify American research was Vannevar Bush. Bush, a mathematician and electrical engineer, served as head of the Office of Scientific Research and Development from 1941 to 1946. Among his duties were coordinating the Manhattan Project and serving as President Roosevelt's science adviser. Based on his wartime experience, in 1945 he wrote an influential report, "Science: The Endless Frontier," which proposed that the government vastly increase its support of basic research. Bush wrote:

---

[16] A. Hunter Dupree, "Influence of the Past: An Interpretation of Recent Development in the Context of 200 Years of History," *The Annals* of the American Academy of Political and Social Science, January 1960, p. 24.

[17] See "Balance of Terror: 25 Years After Alamogordo," *E.R.R.*, 1970 Vol. II, pp. 501-503.

Basic research leads to new knowledge. It provides scientific capital. It creates the fund from which the practical applications of knowledge must be drawn.... A nation which depends upon others for its new basic scientific knowledge will be slow in its industrial progress and weak in its competitive position in world trade, regardless of its mechanical skill."[18]

Bush urged the establishment of a national science board to advise the government on policy related to science and a national research foundation to fund research. The foundation would give money to researchers working in nonprofit institutions such as colleges and universities. The grants would be based on the merits of the specific projects proposed — unlike research funds administered by the U.S. Department of Agriculture, which were based on geographic formulas.

Bush's plan became the outline of the National Science Foundation, created in 1950 after five years of controversy over whether grants should be based on merit and whether scientists should control the money. As the controversy over creation of the NSF was occurring, large segments of the nation's research program were being organized in other federal agencies. The Atomic Energy Commission was established in 1946, with a heavy research emphasis. The National Institutes of Health, a small agency with large ambitions, took over wartime medical research after 1945 and began to grow very rapidly.

NIH, in particular, benefited during the 1950s and 1960s from a Congress that viewed medical research as a winning issue with constituents. Congress routinely added increases to the agency's budget far in excess of what the executive branch had proposed. Within NIH there were created by Congress a number of institutes, such as the National Cancer Institute, the National Allergy Institute, the National Arthritis Institute.

However, the research that each institute was doing internally and supporting in colleges and universities varied little from institute to institute. Historian Stephen Strickland wrote that "the agency could not help but yield to its friends' emphasis on disease problems and solutions rather than continue in its historic preference the exploration of basic science, its self-determined research goals."[19] NIH yielded, however, in name only. In actuality, the emphasis continued to be basic, self-determined research.

Sputnik's Spur to U.S. Science Funding

In 1957 the Soviet Union sent a small unmanned satellite into orbit around the earth and called it Sputnik. That dramatic

---

[18] Vannevar Bush, "Science: The Endless Frontier," U.S. Government Printing Office, 1945, pp. 13-14.

[19] Stephen P. Strickland, *Politics, Science and Dread Disease* (1972), p. 193.

15

event jolted the U.S. government and the American science community into action. Viewing the Soviet space program as a direct threat to national prestige and security, the government began pouring funds into the sciences, both to finance research and to increase science education from the first grade to post-doctoral level. Philip Abelson recently described those years: "Following Sputnik, enrollments increased, large numbers of graduate fellowships were made available, research grants were readily forthcoming, and teaching loads were diminished."[20]

The space race with Russia got going in a big way in the 1960s, culminating in the greatest engineering feat of all time and an American triumph — the placing of a man on the moon. The National Aeronautics and Space Administration, a major scientific agency formed in 1958, had become a fertile source of funds for research and development. Science by now was a growth industry of huge proportions. But the spectacular successes in the skies — and the billions being spent on space exploits — were already being questioned in Congress, by the public and even the scientific community well before the first manned moon landing in July 1969.[21]

There were even earlier cracks in the growing monolith of science. As early as the Manhattan Project, there were doubts on the part of some scientists that the atom bomb was a proper application of science. Leo Szilard, a Hungarian-born physicist who had alerted the Roosevelt administration to the possibilities of building an atomic bomb in 1941, became concerned in 1945 that the bomb would lead to worldwide holocaust. As he had in 1941, Szilard used Albert Einstein as his messenger to the White House. Einstein wrote to Roosevelt asking that he meet with the scientists to discuss the issues. But Roosevelt died before seeing the letter. President Truman, upon succeeding Roosevelt, made a decision to drop the bomb and usher in the age of the atom.

In 1946 a group of atomic scientists, many of whom had worked on the Manhattan Project, formed a group to work for nuclear disarmament and civilian control of nuclear weapons. They called themselves the Federation of Atomic Scientists, but in 1947 changed their name to the Federation of American Scientists, and broadened their goals to all issues affecting relations between science and society. The FAS now calls itself "the conscience of the scientific community."

The FAS represents the first case of what has come to be a rather common phenomenon: scientists calling public attention to problems or dangers flowing from science. It was a scientist,

---

[20] Philip Abelson, "Employment Opportunities for Scientists," *Science,* May 12, 1978, p. 609.
[21] See Congressional Quarterly's *Congress and the Nation,* Vol. II (1969), pp. 531-551.

biologist Rachel Carson, who first publicized the dangers of DDT and other chlorinated hydrocarbon pesticides. Scientists alerted the public to the potential dangers to the ozone layer protecting the earth.[22] Scientists have been prominent in the anti-nuclear energy campaign. Scientists first went public with concerns about recombinant DNA.

Vast public funding for science, combined with the tendency of scientists to take political stands, has inevitably led to demands for more public participation in science. This has been a key issue, for instance, in the debate over recombinant DNA. Barbara Culliton, editor of the "News and Comment" section of *Science,* wrote recently: "The once widespread feeling that scientists alone should have domain over the scientific enterprise is being replaced by a philosophy that calls for public involvement in science, irrespective of the fact that many of the elders of science find the idea abhorrent."[23]

Approaches to Reconciliation

THE CONFLICT between basic research and practical politics may be a matter of boundaries. Harvey Brooks, the Benjamin Pierce Professor of Technology and Public Policy at Harvard, wrote: "By now most politicians concede the necessity for some degree of self-governance and internal agenda setting in the scientific community, and most scientists concede the necessity for some political and public input to the setting of the general directions and goals of scientific research. The issues are where the lines should be drawn...."[24]

Some scholars suggest that the intellectual climate in which those issues will be debated has changed significantly. Philosopher Theodore Roszak has said that the conflict between science and religion of the 17th century was decided in favor of science, which Roszak termed "reason." Reason then became a new, though flawed, religion: "Reason...is the god-word of a specific and highly impassioned ideology handed down to us from our ancestors of the Enlightenment as part of a total cultural and political program. Tied to that ideology is an aggressive dedication to the urban-industrialization of the world and to the scientist's universe as the only sane reality."[25] Roszak

[22] See "Ozone Controversy," *E.R.R.,* 1976 Vol. I, pp. 205-244.

[23] Barbara J. Culliton, "Science's Restive Public," *Daedalus,* spring 1978, p. 147.

[24] Harvey Brooks, "The Problems of Research Priorities," *Daedalus,* spring 1978, p. 178.

[25] Theodore Roszak, *Where the Wasteland Ends: Politics and Transcendence in Postindustrial Society* (1972), p. 171.

said he believed the scientific ideology was breaking down and new forms of belief, which he called "Gnosis" — visionary and ecstatic — was emerging.

Gerald Holton, a science historian at Harvard, has said the West is moving away from an ideology of progress and toward an ideology of limits "that goes much beyond limits to scientific inquiry." Holton connects the debate over greater control of science to the "general new awareness of the existence of or necessity for limits...."[26] The issue of limiting scientific inquiry will be subject to debate for years. Even assuming agreement on the need for limits of some sort and for public involvement in science, hard questions remain: Who will set the limits? How will the democratic impulse of public participation fit with the autocratic need for decisions by trained specialists? How can creativity be preserved in the face of bureaucratic requirements?

The Cambridge Model for Local Control

Many of those questions have already come up in discussion related to regulating recombinant DNA in Cambridge, Mass. Cambridge is significant not just in relation to controlling research on recombinant DNA, which may turn out to be a unique case. It does not appear that other localities are much interested in the DNA issue at the moment, though interest may revive. But the course of events in Cambridge does provide precedents as other cases develop in the control of science.

In August 1976, Cambridge Mayor Alfred Vellucci created a citizens' board — there were no biological scientists among the board members — to consider whether recombinant DNA research being conducted at Harvard could harm the public. Its task was to review the safety procedures spelled out in the NIH guidelines for recombinant DNA research and determine if they were adequate. The issue, as the Cambridge Experimentation Review Board saw it, was who should control research. The board stated that "decisions regarding the appropriate course between the risks and benefits of potentially dangerous scientific inquiry must not be adjudicated within the inner circles of the scientific establishment."[27]

After hearing testimony from supporters and opponents of the research, the board recommended that the research go ahead, with some slight additions to the requirements in the NIH guidelines. The Cambridge City Council endorsed the board's recommendations on Feb. 7, 1977. Stanley Jones, staff director of the Senate Health Subcommittee, said that the action in

---

[26] Gerald Holton, "Epilog to the Issue, 'Limits of Scientific Inquiry' " *Daedalus,* spring 1978, p. 232.
[27] Cambridge Experimentation Review Board, "Guidelines for the Use of Recombinant DNA Molecule Technology in the City of Cambridge," Dec. 21, 1976.

Cambridge was significant because "it is the first time a public community group has looked at an issue in science and made recommendations on what it thought was appropriate."[28]

Although most scientists gave the Cambridge review board high marks, many fear that the concept of local control will spread. The most difficult issue facing Congress as it tries to write legislation governing recombinant DNA research is the issue of local control. Those who are pressing for restrictive legislation want to permit other communities to take the same action Cambridge did. Scientists supporting DNA research, on the other hand, seek legislation to keep localities from regulating the work, much as the federal government in the Atomic Energy Act of 1946 preempts the regulation of nuclear energy.

Science Courts for Resolving Disputes

Some scientists have proposed the creation of science courts to resolve public-policy disputes in which scientists on both sides of an issue assert that their view is the only scientifically correct one. The science court concept is largely the brainchild of Arthur Kantrowitz, chairman of the Avco Everett Research Laboratory Inc., in Everett, Mass. Kantrowitz first raised the idea a decade ago and got no response. But, using his position on several government advisory committees, he began lobbying for a test of his idea. In August 1976 Kantrowitz produced a report explaining how the court would work.[29]

He envisioned that disputes over technical issues — such as safety of nuclear power, use of food additives, or recombinant DNA — would be argued in hearings before a panel of scientist-judges. The hearing would be adversarial, conducted much as a trial is in a courtroom. The goal would be to force each side to confront the other's arguments directly. The judges would be neutral experts in fields related to the issue in dispute. After hearing the evidence, they would prepare a report, noting points of agreement and disagreement and reaching judgment on disputed areas. Many have expressed support for the idea in principle. Others fear that decisions may be awarded on the basis of oratorical skills rather than scientific merit.

It is clear that there will be many proposals in the coming years directed at reconciling the public and scientific viewpoints. Most observers agree that the task ahead is to balance scientific independence and political concerns so that all are served. Science has always been a political institution and nothing suggests that changing that condition is either likely or even particularly desirable.

---

[28] Quoted in *The Christian Science Monitor*, Jan. 17, 1977.

[29] See "The Science Court: An Interim Report," *Science*, Aug. 20, 1976, pp. 653-656.

Books

Blissett, Marlan, *Politics in Science,* Little, Brown, 1972.
Boffey, Phillip M., *The Brain Bank of America,* McGraw-Hill, 1975.
Boorstin, Daniel J., *The Americans: The Democratic Experience,* Random House, 1973.
Bronowski, J. and Bruce Mazlich, *The Western Intellectual Tradition,* Harper & Row, 1960.
Commoner, Barry, *Science & Survival,* Viking, 1963.
Joseph I. Lieberman, *The Scorpion and the Tarantula,* Houghton Mifflin, 1970.
Roszak, Theodore, *The Making of a Counter Culture: Reflections on the Technocratic Society,* Doubleday, 1969.
——*Where the Wasteland Ends: Politics and Transcendence in Postindustrial Society,* Doubleday, 1972.
Strickland, Stephen P., *Politics, Science, and Dread Disease,* Harvard University Press, 1972.

Articles

Dupree, A. Hunter, "Influence of the Past: An Interpretation of Recent Development in the Context of 200 Years of History," *Annals* of the American Academy of Political and Social Science, January 1960.
Daedalus, spring 1978 issue.
Fields, Cheryl M., "Who Should Control Recombinant DNA," *The Chronicle of Higher Education,* March 21, 1977.
Greenberg, Daniel S., "Scientists Wanted — Pioneers Needn't Apply; Call AD 2000," *Smithsonian,* July 1976.
Huebner, Albert L., "The No-Win War Against Cancer," *The Progressive,* October 1977.
Rosenfeld, Albert, "When Man Becomes As God," *Saturday Review,* Dec. 12, 1977.
Science, selected issues.
Thomas, Lewis, "Notes of a Biology-Watcher, The Hazards of Science," *The New England Journal of Medicine,* Feb. 10, 1977.

Reports and Studies

Bush, Vannevar, "Science: The Endless Frontier," Government Printing Office, 1945.
Editorial Research Reports: "Genetic Research," 1977 Vol. II, p. 223; "National Science Policy," 1973 Vol. II, p. 271; "Ozone Controversy," 1976 Vol. I, p. 205; "Quest for Cancer Control," 1974 Vol. II, p. 623.
National Science Foundation, "Science Indicators, 1974."
——"Science Indicators, 1976."
——"Science at the Bicentennial."

Brain Research

by

Philip J. Hilts

**Sept. 15
1 9 7 8**

BRAIN RESEARCH

ONE OF THE more obscure sciences, brain research, has in the past few years become one of the most glamorous. A number of advances in research have not only provided a new understanding of how the brain functions but have started a rain of predictions about the future. "We are on the edge of a choose-your-mood society," a scientist was quoted as saying in the pages of *Fortune* magazine. "Those of us who work in the field [of brain research] see a developing potential for nearly total control of human emotion status, mental functioning, the will to act."[1]

In a similar vein, an editor of *Science News* said last year: "We may well be on the threshold of a provocative new era of using the brain's own chemicals to improve mind and behavior. The discoveries now being made hold exciting and profound implications for helping people suffering from all sorts of behavioral problems...."[2] And in *Quest* magazine, Jo Durden-Smith wrote: "We are at...'the newest frontier.' A new generation of opiates and mood drugs (to be used or abused as future generations see fit), a new control over mental disorder, a new explanation for (and perhaps control over) the activities central to the functioning of human beings — these things seem assured."[3]

The popular magazines have had an outpouring of stories on the subject, and the predictions go on. But even the neuroscientists themselves, who are slower to reach conclusions, are willing to say that the sudden surge forward in the brain sciences may be greater in its effect than the "antibiotics revolution" which has already transformed medicine in this century.

Though decades of research prepared the way, it was only in recent years that the cascade of discoveries and predictions began. The chief discovery was about pain and opiates. Pain, a universal affliction of men and animals, perhaps has no other natural equivalent in its ability to cause reactions of avoidance and fear. It is a powerful motivator and is used by the body to warn of danger, by parents to discipline their children, and by society to curb misbehavior. The lengths to which individuals will go to stop pain are extraordinary — and in the case of abuses

---

[1] Gene Bylinsky, "A Preview of the 'Choose Your Mood' Society," *Fortune*, March 1977, p. 220. The scientist was unidentified.
[2] Joan Arehart-Treichel, writing in *Human Behavior* magazine, March 1977, p. 56.
[3] Jo Durden-Smith, "A Chemical Cure for Madness?" *Quest*, May-June 1978, p. 38.

of narcotics and tranquilizers — sometimes quite costly to society.

Discovery That Brain Has Own Opiates

It has been known for at least 2,500 years that opiates are the most powerful drugs to relieve pain. But it was not until 1973 that Candace Pert and Solomon Snyder at Johns Hopkins University first discovered that the whole family of opiates — heroin, morphine, opium — are merely imitators of the real painkillers. The real opiates are those made by the body itself, to use on itself in critical moments of stress and injury. Pert, a young graduate student in pharmacology working under Dr. Snyder's direction, found that the brain contains specific "receptors" where the molecules of the opiates attach themselves to exert their pleasurable and notorious effects. It is by binding to these receptors on the outside membrane of some brain cells that the opiates work.[4]

In further searching, Pert also found that these opiate receptors were in animals other than humans. In fact, she tracked down specific opiate receptors even in the lowly hagfish, the oldest known vertebrate and a creature which has swum the ocean virtually unchanged by evolution for hundreds of millions of years. Obviously the hagfish and other vertebrates could not be carrying receptors in their brains for the brown dust made from a poppy plant. But the opposite thought set off a race among scientists: if there was a specific opiate receptor, then the body must make its own opiates.

While Pert and Snyder had found the brain's own receptors for opiates, they had not found the brain's opiates themselves. The race to find these natural painkilling chemicals ended in 1975, when John Hughes and Hans Kosterlitz in Scotland detected the substance in pig brains. Their discovery was followed quickly by similar discoveries in other labs, including the lab of Pert and Snyder. The first natural painkillers were called *enkephalins*, from the Greek word for brain. But later, larger versions of the molecules were found that were more effective — 40 times more powerful than the enkephalins and 100 times more powerful than morphine. These new versions are called *endorphins*, meaning "the morphine within."

It is from this one discovery — of the body's own drugs for pain — that much of the excitement about brain research in the past few years has issued. It focuses on the possibility of synthesizing a substance, perhaps a completely natural and non-addictive substance, which is dozens or hundreds of times more powerful

---

[4] See Solomon Snyder's account of the discovery, and its background, in his article "The Body's Natural Opiates," in the *Encyclopaedia Britannica's 1979 Yearbook of Science and the Future*, pp. 103-115.

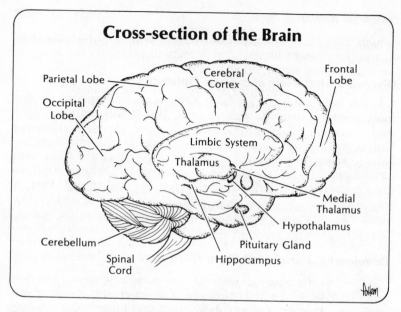

Cross-section of the Brain

Parietal Lobe

Cerebral Cortex

Frontal Lobe

Occipital Lobe

Limbic System

Thalamus

Medial Thalamus

Hypothalamus

Cerebellum

Pituitary Gland

Spinal Cord

Hippocampus

than morphine. "A non-addicting opiate drug would be a miraculous benefit," said Dr. Snyder, "in treating both severe, chronic pain and a variety of emotional problems." Richard Restak, a neuroscientist, points out it is possible to imagine saving millions of people from drug addiction and possible to see this one thing alone as having effects throughout society, from reducing crime rates to improving international relations. The politics of drugs have influence on the diplomacy conducted with several nations, such as Turkey and Mexico.[5]

The discovery of the body's own opiates immediately suggested explanations for several previously puzzling phenomena. Barbara Villet, a medical writer, recounted recently in *Atlantic* magazine that there have long been clues to the presence of endorphins in the brain and body. "Wartime accounts of cruelly wounded men whose indifference to their own agonies often puzzled front line observers are now thought by some neuroscientists to represent 'case studies' of an endorphin response to trauma, and there is speculation that religious trance states which permit individuals to walk on white hot coals or pierce their flesh with impunity may demonstrate the analgesic actions of these substances as well."[6]

The way the opiates have their effect is suggested by the placement of the body's special "opiate receptors." They are primarily in two spots: (1) the *limbic* part of the brain, which is the center of most strong emotions, such as fear, rage, love, depression; and (2) the *medial thalamus,* the way station at the base of the brain

[5] Richard Restak, "The Brain Makes Its Own Narcotics," *Saturday Review,* March 5, 1977. See also *The Politics of Drugs* (1975) by Richard C. Schroeder.
[6] Barbara Villet, "Opiates of the Mind," *Atlantic,* June 1978, p. 82.

which transmits pain impulses from the body to the brain. Thus, the opiates might interfere with pain signals from the body, as well as having strong effects on emotions.

Currently the prevailing theory about addiction and the endorphins is one that has proved true about other natural substances. When an artificial endorphin (heroin, for example) begins binding itself to the receptors on brain cells, there is a natural feedback system which becomes fooled. The message goes to the site of manufacture of endorphins, and it says: "The cells have enough endorphin, stop production." When natural production of endorphins deteriorates and heroin is also stopped, then the system has a sudden shortage of endorphin. The system needs more, fast, and cannot make its own because the endorphin factories are shut down. But heroin will bring relief.

Developing Theories of Brain Chemistry

After the discovery of endorphins, other discoveries followed quickly. Earlier research began to take on greater meaning and usefulness, and as with any dawning of new understanding, rows of scattered facts and curious pieces of data began to fit into place. This stage is then soon followed by a whole new set of questions, as complicated as the first but at a higher level.

One popular theory is that the related brain chemicals are all part of a general "crisis system." They are the means by which the "fight or flight" response takes over the body. In such sudden stress, the body must prepare itself for peak performance — decreasing sensitivity to pain, increasing strength, heightening perception, and strengthening emotional motivators. This crisis response occurs suddenly as if a chemical were injected into the system.

It is now being proposed that perhaps one chemical, pro-opiocortin, is released under stress. The giant molecule of this chemical then breaks up into smaller, behaviorally active components — enkephalin, endorphins, beta-lipotropin, and ACTH (adrenocorticotrophic hormone), which are all part of the same amino acid chain. Endorphins and ACTH are both linked to stress; they seem to cause several reactions useful in stress such as providing improved visual perception and pain relief. Sidney Udenfriend, the discoverer of pro-opiocortin, uses a military analogy for this action. "[L]ike a MIRV guided missile, with multiple independently targetable warheads, the precursor molecule breaks up into small pieces — ACTH, MSH, endotrophins, enkephalins, each homing in on its own target issue."[7]

Some recent studies have reported significant results without

---

[7] Quoted in *Medical World News*, Jan. 9, 1978, p. 96.

necessarily suggesting any clear mechanism by which these successes were achieved. Candace Pert (now a Ph.D.) and her husband, psychobiologist Agu Pert, together with their co-worker John Tillman, recently reported finding another natural opiate circulating in the blood rather than within the brain alone; it is called *anondynin*. The Perts are cautiously speculating that anodynin may be a "healing hormone" important in stress, pain, and sleep. More important, they think it may play a role in drug addiction.

Dr. Floyd Bloom, a Nobel laureate, and his colleagues at the Center for Behavioral Neurobiology of the Salk Institute, at La Jolla, Calif., injected an endorphin (beta-endorphin)[8] directly into the cerebrospinal fluid of several rats. After 30 minutes the rats became immobile and catatonic, "frozen into postures so rigid they could be suspended across bookends." But within seconds after an injection of a drug that is a direct antagonist of the opiates, the rats recovered completely. This clue, and others, indicates that some of the symptoms seen in mental illness may be caused by derangements of the body's mechanisms regulating endorphins.

Carrying the work further, Bloom found that the beta-endorphin they were using could be split into two other active compounds — alpha and gamma endorphins. Though alpha and beta differed by a single amino acid,[9] they produced opposite effects on the rats. Alpha endorphin produced pain relief while gamma endorphin made the animals violent and apparently overly sensitive to pain.

Effect of Chemicals on Memory, Learning

Dr. David de Wied, a Dutch pharmacologist, began a series of impressive experiments in the late 1950s on the chemistry of learning and memory. Other experimenters had noted the effects of some chemicals on memory. For example, Dr. Bernard Agranoff of the University of Michigan found he could inflict memory loss on goldfish by injecting puromycin into their brains. Puromycin is an antibiotic that blocks the formation of protein. He injected the drug after the goldfish had learned a new skill. The goldfish carried the memory for a short time, but were unable to "fix" it chemically to make it a long-term memory.

Dr. James McGaugh of the University of California at Irvine later found that minute doses of strychnine boosted the ability of rats to learn and remember. Under normal conditions, a rat could learn the turns in McGaugh's maze in about 25 trials. With

---

[8] What came to be named beta-endorphin was a protein identified and isolated from the tissue of a sheep's brain in 1964 by Dr. C. H. Li, director of the Hormone Research Laboratory at the University of California at San Francisco.

[9] Endorphins are chains of amino acids; beta lipotropin, the parent molecule, has 91 amino acids in its chain.

the proper dose of strychnine, however, the rats managed to learn the maze in half a dozen trials. One of the interesting lessons of this experiment was that the drugs which affect memory were far more effective with the older rats than with the younger ones.

In De Wied's work, the pituitary gland of a group of rats was removed. It was found that this surgery prevented the rats from learning how to respond to such stimuli as lights, buzzers and mild shocks. But after De Wied injected the rats with ACTH, a larger relative of the endorphins, he found that their ability to learn had been restored, despite the fact that they were missing pituitary glands. In 1965, De Wied tried a similar experiment with MSH (melanocyte stimulating hormone), another chemical cousin to the endorphins.

These experiments with ACTH and MSH led him down an important line of thought several years in advance of the endorphin discoveries. ACTH and MSH were well known hormones whose effects were believed to be simply those of controlling pigmentation (in the case of MSH) and releasing steroids during stress (in the case of ACTH). But when De Wied found they had impressive behavioral effects, be began to look for other pituitary hormones which might have unrecognized effects on the brain.

In this search, he found vasopressin. This tiny protein was thought to be merely a regulator of the body's water and salt retention. But De Wied learned that it was the most powerful memory drug yet tested. Further experiments in the past decade have demonstrated its effect in humans — those who are normal, those who are mentally retarded and those who are afflicted with senility. All who received added amounts of vasopressin showed significant improvement in visual attention and memory.

Endorphin Testing With Mental Patients

Like other chemicals in the beta-lipotropin family, MSH was found to have other positive effects on behavior. For example, a group of researchers in New Orleans, including Abba J. Kastin, Lyle Miller and Curt Sandman, discovered that MIF (MSH-release inhibiting factor), a substance which may control the release of MSH, had a positive effect on the tremors and rigidity of patients with Parkinson's disease. But when given in conjunction with L-dopa, the drug currently used for the disease's treatment, the combination virtually abolishes the symptoms. Small doses of MIF can also counter depression, Kastin found. So, too, can other protein, TRH (thyrotropin releasing hormone).

Hearing about Dr. Bloom's work at La Jolla injecting rats with synthesized beta-endorphin (see p. 27), the following year (1977) two psychiatrists, Drs. Nathan Kline of the Rockland

Research Institute at Orangeburg, N.Y., and Heinz E. Lehmann of Montreal, took the next step: They injected a small amount of the chemical into humans. They selected from Kline's private practice 14 patients with a variety of severe mental illnesses who had not responded to any of the usual drugs and therapies. Injections were also given to another person who was not a mental patient.

Kline and Lehmann gave a preliminary report of their work at a conference of the American Psychiatric Association in 1977 and a fuller report at a symposium held in San Juan, Puerto Rico, at the end of the year. They told their colleagues that seven of the patients showed immediate improvement in mood and behavior. These findings were not received uncritically, however, because only a few patients were tested and the testing was not "controlled." Controlled testing commonly requires that it be "double blind" — that neither the patients nor the testers know which of the patients are receiving actual medication and which are receiving useless placebo pills.

Others who had meanwhile used endorphins to treat mental illness also reported on their work at the San Juan meeting. Dr. Jules Angst of the University Psychiatric Clinic in Zurich said his experiments with six patients were somewhat positive, though not conclusive, and generally "the hypothesis is justified that beta-endorphin may have anti-depressive properties and potential to switch depression to hypomania (a mild form of mania) and mania." Two Swedish scientists, Drs. Agneta Wahlström and Lars Terenius, said they had found another link between endorphins and mental illness when they found a version of endorphins in the spinal fluid of schizophrenics.

Wahlström added a note of skepticism, however, by saying that findings of two years earlier (of endorphins in spinal fluid) had turned out not to be as she had reported them. The chemicals actually found were endorphins "distinct from the known ones," she said at San Juan. There were also indications that the work on hallucinations reported earlier by the Swedish team had not been repeated in strictly controlled experiments.[10]

Experiments using endorphins in the treatment of mental illness still go forward, though with less certainty and perhaps with less enthusiasm than immediately after the initial discoveries. The list of studies with startling or curious results in many areas of brain research is long. Laboratory tests suggest the possibility that injections of brain proteins may do such varied things as quickly produce sleep and restore sexual potency.

---

[10] Wahlström and Terenius, together with Swedish colleague Lars-Mangus Gunne, reported in 1975 that the hallucinations of schizophrenics could be halted by naloxone, a drug that locks onto the opiate receptors like endorphin. For an account of the San Juan meeting, and its background, see David Leff's reporting in *Medical World News*, Jan. 9, 1978, pp. 86-96.

Naturally, there is skepticism about the cure-all quality of the newly discovered brain chemicals. But it is certain that more of them will be turned up, since those that have been discovered so far mostly affect only the limbic area of the brain. Many other areas are yet to be explored.

Biologic Basis of New Advances

THERE IS a unique quality about brain research that sets it apart from other forms of research. The human brain is the only thing known to mankind which attempts to understand itself. It is self-awareness extended to an extraordinary degree. Santiago Ramon y Cajal, a great microscopist who was the first to draw accurate maps of brain cells, said more than half a century ago that knowledge of the physical and chemical basis of memory, feelings, and reason would make man the true master of creation, and this his most transcendental accomplishment would be the conquering of his own brain.[11]

Such a view might well have been inconceivable to the ancients. It was not clear then that all these human properties inhabited the same place, much less that the place was the brain. Aristotle believed that the brain was just a mechanism for cooling the blood; the heart, he felt, was the seat of reason and thought. About five centuries later, the Greek physician Claudius Galen named the brain as the center of thinking, but he had no tools to investigate it or means to understand it.

Human understanding of the brain did not move forward from the time of Galen until practically our own time. Descartes (1596-1650) was among the first to demystify the workings of the body and brain. He imagined the body as no more than complex, wonderful machinery, but then in order not to offend the Church added that the whole was animated by the soul, which had its residence in the brain's tiny pineal gland.

It was discovered early in the 19th century that electricity coursed through the body and brain, and actually seemed to cause bodily movement. This idea, besides giving rise to Mary Shelley's classic novel *Frankenstein,* led to early experiments in the electrical stimulation of the brain. It was demonstrated in 1870 that stimulating a dog's brain would cause the animal's limbs to move. The "vital spirits" necessary to explain action in earlier times soon became useless appendages which created much confusion among philosophers.

[11] See José Delgado's *Physical Control of the Mind* (1969).

Some of the confusion over mind and matter was reflected in the work of Sigmund Freud. While maintaining that the mind had a purely physical basis, he nevertheless ended up founding a school of psychology whose investigations led in a different direction — toward the analyst's couch. It is interesting that Freud was one of the earliest and strongest advocates of the use of drugs as agents of behavioral therapy. He took cocaine himself, found its properties to be "magical," and then began to press the drug on his sisters, colleagues and patients. Indeed, it has been said he introduced to Europe "the third scourge of humanity" (alcohol and morphine were the others). But deaths began to result from his enthusiasm — one patient and a close friend died of the drug — and finally Freud stopped using it.

Freud's use of the drug did, however, certify and reinforce the idea that drugs offered one method of helping to understand and treat the problems of the mind. Parodoxically, Freud's own psychoanalytic movement, and therefore much of the body of psychological study, did not move in that direction. Rather its work distracted many from the biological basis of behavior and concentrated attention on the abstract constructs which were thought to make up the mind.

Electric Stimulation, Shock and Surgery

Meanwhile, the progress of biological investigations of the brain was slow. Electrical stimulation of the brain demonstrated in 1932 that a variety of body movements and emotional reactions could be sparked from the brain. But between that time and this decade, apart from refining this information and perfecting the tools to invoke it, little more was done to further the electrical control of the mind.

One monumental investigation of the brain during those years was the work of Karl Lashley. It was believed that memory was an electrical property of the brain, that it was made up of certain connections among the brain's circuitry. Lashley called these lasting patterns of memory "engrams," and he spent his life looking for them. Convinced that they were located somewhere on the convoluted cortex of the brain, he trained thousands of rats to run mazes and learn other skills. He then carefully and systematically cut off bit after bit of cortex. At some point, he expected that he would make a cut that would erase the rat's memory of his training. But cut as he would, the memory remained.

"Even when the brain injury was severe enough to make the animal limp or stagger," writes Maya Pines, "even when 90 percent of its visual cortex had been removed, the rat still found its way across the maze. Finally, after decades of struggling with the problem of how information is encoded in the brain and how people learn, Lashley came to the tongue-in-cheek conclusion: "Learning is just not possible at all."[12]

In recent years, other types of brain research involving electrical stimulation have attracted scientific and public attention.[13] Where it is medically necessary to destroy brain tissue, the surgical procedure in use today is to stimulate the brain electrically to elicit the sort of behavior that is troublesome and then electrically burn a tiny — practically invisible — lesion into the brain tissue. This is a far cry from the earliest form of psychosurgery, the simple lobotomy, a surgical operation destroying sizable amounts of the brain.

The first known pyschosurgery can be traced to 1891 but its great popularity was achieved between 1935 and 1955, after the Portuguese surgeon Egaz Moniz invented the technique of prefrontal lobotomy — cutting out the frontmost areas of the brain, where the higher mental activities are believed to occur, in order to relieve extreme anxiety. Lobotomies eventually became less and less popular when it was clear that they treated illness by creating a zombie-like state called "frontal lobe syndrome."

Until the widespread use of drugs as agents of behavioral therapy, psychosurgery and electric shock were for decades the two most common forms of treating severely disturbed mental patients. Electroshock therapy, unlike psychosurgery, is still used today despite the considerable criticism it has received in recent years. It is estimated that between 50,000 and 200,000 patients receive shock treatment each year. Memory loss is not unusual but the severe convulsions which used to be one of the undesirable side effects can now be prevented by a drug injection. Some medical authorities consider shock therapy the single most effective treatment for major depression. However, as with so many other approaches to the treatment of mental illness, the beneficial effects tend to be temporary and do not "cure" the illness.

Beginning of Drug Treatments in 1950s

Drug therapies are now considered more promising and more useful than either shock treatment or psychosurgery. However, both electrical and surgical techniques used in animals have become important in other kinds of brain research. The important value of drugs as agents of behavior therapy was rediscovered — not by psychiatrists this time, but accidentally by doctors.

---

[12] Maya Pines, *The Brain Changers* (1975).
[13] See "Human Engineering," *E.R.R.*, 1971 Vol. II, pp. 376-378.

Surgeons using the chemical chlorpromazine on their patients to help reduce secretions during operations found the patients were calmer than those who had not been given the drug. This observation led to the use of chlorpromazine as a treatment for mental disorder. A similar circumstance in the treatment of tuberculosis led to the discovery of MAOI (monoamine oxidase inhibitor) as a treatment for depression and lithium as a treatment for mania.

These drug treatments were first introduced into mental hospitals early in the 1950s. The effects were startling. Before drugs were used to treat schizophrenia, depression and mania, the number of patients in mental institutions had grown rapidly. After World War I, there were fewer than 200,000 mental patients in public care; by 1956, the number approached 560,000. Citing those figures, Philip Berger wrote in *Science* magazine this year:

> Firsthand descriptions by physicians who worked with the mentally ill before the introduction of effective pharmacotherapies paint a dismal picture of the pre-drug era.... Most patients were sent quickly to state mental hospitals, which were more like custodial facilities than medical treatment centers. Pessimism about psychiatric disorders was widespread, admissions increased, and discharges remained low.[14]

Since that time, the number of patients has fallen below 200,000, and the number is still declining. Doctors say that the character of mental institutions has changed for the better, even though glaring defects still remain.

The drug treatments themselves are not ideal, are sometimes dangerous, and do not work in all patients. But they are generally effective. Used to treat schizophrenia are a number of "antipsychotic" chemicals which appear to have a similar effect on receptors in the brain. These particular receptors in normal circumstances are activated only by the body's own neurotransmitter called dopamine. The anti-psychotic drugs apparently work by damming up these dopamine receptors so that the dopamine cannot stimulate the nerve cells.

Drugs frequently used for affective illness, or depression as it is more commonly called, are tricyclics. For many patients, tricyclics improve the mood of depressed patients, restore confidence, eliminate suicidal thinking, and relieve many physical complaints as well. It is believed that tricyclics work by increasing the action of the important neurotransmitters norepinephrine and serotonin.

For mania, the drug most likely to be used is lithium car-

---

[14] Philip A. Berger, "Medical Treatment of Mental Illness," *Science*, May 26, 1978, p. 974.

Bodily Rhythms and Psychic Ills

Although brain chemicals have been heralded recently, it is possible that other discoveries of the past two decades might be more important in the long run. One discovery has led to a new field called chronobiology or circadian cybernetics. Researchers have found that not only the whole body, but individual systems and cells in the body, all operate in strict time schedules.

These rhythms operate on 24-hour cycles, they take the name circadian rhythms from the Latin words *circa dies* — "about a day." When these rhythms are disrupted, catastrophic change can result. One mild example of disrupted circadian rhythms is "jet lag" — when otherwise healthy individuals become quite ill or disoriented by swiftly changing time zones. Since even down to the level of single cells, animals show strong circadian patterns of behavior, many important functions of the body can be disrupted by being knocked out of phase with the rest of the system.

Dr. Thomas Wehr, chief of the clinical research unit in the psychobiology branch at the National Institute of Mental Health, has begun studies to demonstrate that at least some forms of mental illness may be the result of disrupted rhythms within the body. Many mental illnesses show strongly rhythmic patterns, including severe manic-depression which switches between mania and depression in quite regular cycles.

Wehr studies this "switch" mechanism as an example of the importance of circadian rhythms to behavior, and recently has succeeded in bringing a middle-aged woman out of her severe manic-depressive cycle. Wehr monitored several of the woman's circadian rhythms, then designed a rigid schedule of sleep and eating which forced the several circadian rhythms back into step with one another. The woman's long-standing illness was soon relieved. Other experimenters have also begun to report similar results. Another interesting bit of circadian research, devised by Charles Ehret and Van Potter at Argonne National Laboratory near Chicago, is a system that cures jet lag by resetting biological clocks.

"We have tried to show that illnesses can result when the body's temporal cycles are disrupted," said Dr. Wehr. "Just as medicine has understood for years that tissue disorders can cause illness, it looks now as if we are beginning to learn that temporal disorders can do the same. And as there are glands which secrete hormones to regulate action, it looks as if the central 'clock' which keeps the other bodily clocks in line, acts like a gland 'secreting' rhythms instead of hormones."*

* Interview, Aug. 15, 1978.

bonate. It is effective in reducing the number and severity of manic outbursts in many patients, although the way it works is comparatively less known than tricyclics and major tranquilizers. It is possible that, since depression and mania are the face and tail of the same coin, the same brain system is involved in producing them. If too little action of norepinephrine causes depression, then it is possible that too much action of the drug can cause mania. There is some evidence to show that lithium does decrease the action of norepinephrine.

In the two decades since these chemicals were introduced in the treatment of mental illness, a great deal of work has been done on neurotransmitters, receptors, and electrical action of brain cells. The discovery of psychoactive drugs in the 1950s was the catalyst which set off the flurry of research along chemical lines — research which led eventually to the endorphins. "These accidental discoveries led us back to the biological basis of behavior, and that was inevitable," said Richard Restak. "Imagine a patient with heart trouble who went to the doctor about it. Then imagine the doctor saying to the patient, 'I don't know anything about the heart, I don't study that. But I'll treat you anyway.' This is not very different from a psychiatrist who doesn't study the brain, but still treats mental patients. It had to change."[15]

New Knowledge of Chemical Transmission

It is obvious now that previous images of the brain's action as rather dry and electrical, with only a few chemical transmitters between cells, was quite erroneous. The brain cells are afloat in chemical message-transmitters. But work on these chemicals has only begun, and their action is not simple. What is clear, though, is that this whole area of brain research focuses on one item of brain structure: the *synapse*. It is here that all the chemicals carry out their action.

All the abilities of the brain arise from the firing of its ten billion neurons. The neurons pulse continuously, like tiny telegraphers, but all those ten billion neurons firing independent of one another would still come to nothing. It is when they fire in chorus, when a jolt of electrochemical action is passed from cell to cell, in waves of thousands and millions — then the activity becomes interesting. So it is the synapse, the space between brain cells where the message is passed from one cell to the next, that is the critical spot.

"Consciousness, learning, and intelligence are all synapse-dependent," writes neuroscientist Steven Rose. "It is not too strong to say that the evolution of humanity followed the evolution of the synapse."[16] The pulse is passed along the body of the brain cell electrically. But when it comes to crossing the synapse, the message from cell to cell is transmitted chemically. It is the chemical message across the synapse which is most influential in

---

[15] Interview, Aug. 30, 1978.

[16] Steven Rose, *The Conscious Brain* (1975), p. 65. The book provides a full description of current knowledge about neurotransmitter processes.

Endorphins and Acupuncture

One of the interesting side benefits of the discovery of the endorphins is an explanation for the mysterious effect of acupuncture. A Toronto neurologist, Dr. Bruce Pomeranz, applied acupuncture to cats and monitored the activity of the neurons responsible for pain. Twenty minutes after the needles were applied, the pain messages were blocked.

Then, as a test of whether the endorphins might be causing this pain relief, Pomeranz injected the cats with naloxone. This drug binds itself to the same receptors as opiates and endorphins. It forms a stronger bond than either of them but has no pain-relieving properties. So a dose of naloxone should halt the positive action of the acupuncture.

Pomeranz found that the pain impulses started up again immediately after the injection of naloxone. Interestingly, when he placed the acupuncture needles in the wrong sites — places other than the traditional acupuncture spots — the pain was not relieved. Other studies support his findings, and it is now believed that acupuncture works this way: Pain messages from the parts of the body do not cause pain unless there are sufficient numbers of them at once, creating a sort of threshold effect. Acupuncture stimulates the release of endorphins into the nerve synapses, and blocks these pain messages before they can fire and be registered by the brain.

determining how a neuron shall fire. And it is also this chemical process that is most vulnerable to failure.

The end of one cell reaches out for a neighboring cell. The electrical pulse passes down toward the nerve ending, and there somehow causes a chemical to squirt out across the synapse. The chemical which bears the signal across the gap is called the neurotransmitter. Across the synapse at the neighboring cell are the receptors which receive the neurotransmitters as they swim the short lap from cell to cell.

The receptors will not accept any chemical, but only that neurotransmitter that is chemically right for it. When the transmitter chemical and receptor meet, they lock together in a short chemical embrace. For a moment they form a new molecule, and this new situation is enough to alter the electrical polarity of the receptor's cell. Thus, a new electrical pulse is generated and races through the cell to some other, distant synapse, where the exchange occurs again. The messages pass from cell to cell, and the arrays of firing cells make sensation: in one area of the brain the storm of activity may be pain or pleasure, in another area it may be visual images.

Here is the rub: the chemical action across the synapses can easily become fouled. The neurotransmitter chemicals can be

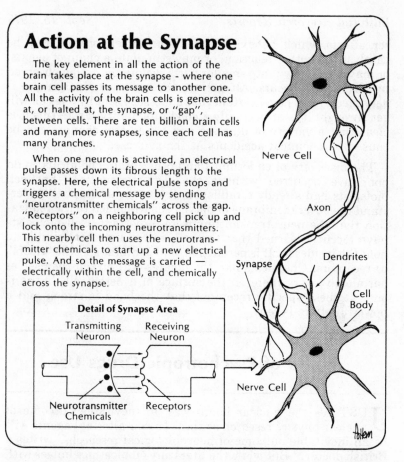

Action at the Synapse

The key element in all the action of the brain takes place at the synapse - where one brain cell passes its message to another one. All the activity of the brain cells is generated at, or halted at, the synapse, or "gap", between cells. There are ten billion brain cells and many more synapses, since each cell has many branches.

When one neuron is activated, an electrical pulse passes down its fibrous length to the synapse. Here, the electrical pulse stops and triggers a chemical message by sending "neurotransmitter chemicals" across the gap. "Receptors" on a neighboring cell pick up and lock onto the incoming neurotransmitters. This nearby cell then uses the neurotransmitter chemicals to start up a new electrical pulse. And so the message is carried — electrically within the cell, and chemically across the synapse.

Nerve Cell

Axon

Synapse

Dendrites

Cell Body

Nerve Cell

Detail of Synapse Area

Transmitting Neuron

Receiving Neuron

Neurotransmitter Chemicals

Receptors

destroyed en route if another chemical is present that reacts with it. The message sent will never be received. Or the transmitter locks onto its proper receptor but remains locked together with it too long; in that event, the passage of future messages will be blocked. Or again, if a foreign chemical similar to the transmitter enters the synapse, it can trigger the action falsely in one cell and thus start up a chain reaction of firings in that area of the brain.

The body itself uses these tricks to control brain action. If neurotransmitter chemicals stay in the vicinity of a receptor site too long, they will cause continuous firing. So the body uses enzymes at the site to destroy the transmitters, and the body also has a mechanism to take transmitter chemicals back into the original cell for reuse later. Either of these methods of inhibiting transmitter chemicals can be ruined by chemical action.

It may take two, three, or more chemicals acting on each other, or on the cell, to carry a single signal. It is this difficulty that creates much of the confusion about which neurotransmitter affects which behavioral problem. An endorphin may be catalyst to another reaction, or may stimulate a chain of events, or may break down into smaller parts, before the action observed by researchers occurs. Altogether, the action at the synapse is very messy. It is never certain that a message sent by one cell will get through; it is at best only a good probability. From this it is clear that a variety of illnesses and behavioral changes can be caused by chemical accidents at the synapse.

The discovery of endorphins and other brain chemicals could not have occurred without the firm foundation that cell biologists had already established. It was the attempt to understand the outer membrane of nerve cells which led to the realization that the neurotransmitters were important. Cell biologists have recently learned that the cell wall is not there simply to hold the insides in. It is permeable, and contains specific "doors" or receptors. Chemicals like the opiates will attach themselves harmlessly everywhere on the surface of a nerve cell, but it is only at these specific receptors that the bond is strong and a powerful effect is felt.

Ethics of Psychotropic Drugs Use

JUST AS the nuclear bomb was an inevitable outcome of nuclear physics research so these [mood-altering] agents will be an inevitable outcome of neurobiological research," writes a British doctor.[17] But while the literature on bioethics bulges with articles and books on psychosurgery and electroconvulsive therapy, the literature on the ethics of psychotropic drugs is surprisingly thin by comparison, even though 10 percent of the American people are users of psychotropic drugs, primarily tranquilizers.

In trying to imagine the problems posed by future "mood drugs" it is useful to look at the current ones offered on the market. Valium was introduced in 1963 and by 1972 some 50 million prescriptions were being written, making it the most frequently prescribed drug in the United States. Librium was introduced in 1960 and by 1972 accounted for 20 million prescriptions.[18] All indications point to increased use of tranquilizing drugs since then.

The boom in sales of these drugs can be laid to several factors,

[17] D. G. Grahame-Smith, "Self-Modification With Mood Changing Drugs," *Journal of Medical Ethics*, 1975, Vol. I, pp. 132-137.
[18] Foregoing figures cited by Ingrid Waldron, in "Increased Prescribing of Valium, Librium, and Other Drugs — An Example of the Influence of Economic and Social Factors on the Practice of Medicine," *International Journal of Health Services*, Vol. 7, No. 1, 1977.

most obviously the desire of patients to take such drugs and the complementary desire of pharmaceutical companies to sell them. The number of drug prescriptions in the United States has more than doubled since 1950, without a commensurate improvement in the nation's health. Deaths due to adverse drug reactions are now about as frequent as deaths from car accidents, Ingrid Waldron reported last year in the *International Journal of Health Services*. Both psychiatrists and laypersons have suggested indiscriminate drug-taking will breed frustration and alienation in the long run because it is a poor substitute for changing the human environment or learning to cope with problems. Dr. Nathan Kline had said early in this decade: "It may be that a certain amount of disruption, disharmony, instability is part of what gives us the thrust to keep moving, so I'm not sure that tranquility is really what one should aim for."[19]

Questions Arising Over Brain Chemicals

It is ironic that the first ethical problem with the use of the new brain chemicals came with their very first trial. Many doctors objected to the manner in which Dr. Kline tested endorphins on his private psychiatric patients *(see p. 29)*. The case also indirectly demonstrates another problem, and perhaps the most serious ethical problem which will be raised by the use of the new brain chemicals. That is the use of drugs on mental patients, either with or against their will. It is believed by some, notably by psychiatrist-author Thomas Szasz,[20] that confining mental patients and drugging them, even with their consent, is criminal behavior. His argument states that the label "mentally ill" says no more than that the individual behaves in ways that are socially unpleasant to those in authority.

The ethical question ultimately becomes: Should we allow people to decide for themselves what state of mind they may be in, provided there is no direct harm to other persons? The choice in this question is not just between freedom and paternalism, for there are degrees of freedom. One of the most troubling aspects of the question is whether anyone can have enough information about psychotropic drugs to make wise choices about their use. Those who consider questions about their use inevitably suggest consideration of some degree of government control of these substances and fuller public information about them.

But there appears to be no suggestion in the popular or scientific literature that the current research should be stopped or restricted. And this may be the clearest comment of all that the expectation is for the positive benefits to outweigh the negative.

---

[19] Spoken at a panel discussion, "Psychotropic Drugs in the Year 2000," cited by Thomas Dugan in a *Hofstra Law Review* article, "The Legal and Social Implications of Psychopharmacology," summer 1974 issue.

[20] His books include *The Myth of Psychiatry* and *The Myth of Mental Health*.

Books

Calder, Nigel, *The Mind of Man,* Compass Books, The Viking Press, 1970.

Delgado, José, *Physical Control of the Mind,* Harper Colophon Books, 1969.

Pines, Maya, *The Brain Changers,* Signet, New American Library, 1975.

Restak, Richard, *Pre-Meditated Man,* The Viking Press, 1973.

Ritchie-Russell, W. and A. J. Dewar, *Explaining the Brain,* Oxford University Press, 1975.

Rose, Steven, *The Conscious Brain,* Alfred A. Knopf, 1975.

Articles

Arehart-Treichel, Joan, "Proteins on the Brain," *Human Behavior,* March 1977.

—— "Brain Proteins, Matter Over Mind," *New York,* April 10, 1978.

Berger, Philip A., "Medical Treatment of Mental Illness," *Science,* May 26, 1978.

Blackwell, Barry, "Psychotropic Drugs in Use Today," *Journal of the American Medical Association,* Sept. 24, 1973.

Bylinsky, Gene, "A Preview of the 'Choose Your Mood' Society," *Fortune,* March 1977.

Chase, Michael, "Secret Life of Neurons," *Psychology Today,* August 1978.

Dugan, Thomas, "The Legal and Social Implications of Psychopharmacology," *Hofstra Law Review,* summer 1974.

Grahame-Smith, D. G., "Self-Modification with Mood-Changing Drugs," *Journal of Medical Ethics,* Vol. I, 1975.

Iversen, L. "Neuropeptides," *Science,* Dec. 23, 1977.

Leff, David, "Doctors Debate Brain Hormone Dilemmas," *Medical World News,* Jan. 9, 1978.

Lennard, Henry, et al., "Hazards Implicit in Prescribing Psychoactive Drugs," *Science,* July 31, 1970.

Pines, Maya, "Speak, Memory," *Saturday Review,* Aug. 9, 1975.

Restak, Richard, "The Brain Makes Its Own Narcotics," *Saturday Review,* March 5, 1977.

Synder, Solomon, "The Body's Natural Opiates," *Encyclopaedia Britannica 1979 Yearbook of Science and the Future.*

Waldron, Ingrid, "Increased Prescribing of Valium, Librium, and Other Drugs — An Example of the Influence of Economic and Social Factors on the Practice of Medicine," *International Journal of Health Services,* Vol. 7, No. 1, 1977.

Reports and Studies

Editorial Research Reports, "Schizophrenia, Medical Enigma," 1972 Vol. I, p. 233; 'Medical Ethics," 1972 Vol. I, p. 461; "Psychomedicine," 1974 Vol. II, p. 497.

The National Commission for the Protection of Human Subjects of Biomedical and Behavioral Research, "Report and Recommendations on Psychosurgery," 1976.

MYSTERIOUS PHENOMENA: THE NEW OBSESSION

by

William V. Thomas

Jan. 20
1 9 7 8

MYSTERIOUS PHENOMENA

B Y ALL INDICATIONS, America is in the midst of a mystical revolution. Inundated with tales of extraordinary phenomena and claims of strange psychic powers, people in increasing numbers are turning to belief in rationally unex-
plainable—paranormal—phenomena. Reincarnation, mind
reading, visitations from outer space and other marvels, once
considered the exclusive province of science fiction, have gained
popular acceptance as fact.

A good illustration of the trend is the growth of interest in
astrology. A Gallup Poll last year indicated that 32 million
adults in the United States believed in astrological predictions.
Whereas a generation ago only 100 daily newspapers published
horoscopes, today some 1,250 do. But those figures may tell only
a part of the story. Further evidence suggests that many
Americans are apparently willing to embrace a great variety of
beliefs dismissed by science and most Western religions as pure
superstition.

Books promoting the paranormal and the occult continue to
be leading sellers,[1] while current films about extraterrestrial
life, "Star Wars," "Close Encounters of the Third Kind,"
appear certain to break movie box-office records. Enrollments
in college courses that deal with these subjects are also running
high. Throughout a broad segment of society, according to
writer Lee Nisbet, the quest for political and social utopias has
been replaced by a desire to enter into "realms of mystical experience and mystical knowledge."[2]

The new fascination with penetrating the unknown includes
exotic ideas and practices, such as psychic healing, extrasensory
perception, astral projection, pyramid power, dianetics, Kirlian
photography and witchcraft. The list is long. All this has led to
expressions of concern that the wave of belief in "otherworldly"
phenomena may pose a threat to the very foundation of reason.
In 1975, 186 of the country's leading scientists and philosophers

---

[1] Currently five books on *The New York Times* best-seller list deal with paranormal or psychic phenomena. Under fiction: *Illusions* by Richard Bach and *The Book of Merlyn* by
Terence H. White. Under nonfiction: *Gnomes* by Wil Huygen, *The Amityville Horror* by Jay
Anson, and *The Second Ring of Power* by Carlos Castaneda.
[2] Lee Nisbet, "Mystical Aspects of Science—An Exorcism," *The Humanist,* May-June
1977, pp. 43-44. *The Humanist* is a publication of the American Humanist Association and
the American Ethical Union.

signed a statement denouncing astrology and other manifestations of what they called "the new nonsense." That the protest received little news coverage, they said, offered yet another example of the public's preference for sensationalism.

Defenders of traditional methods of investigation would like to see theories of the paranormal subjected to the same standards of accuracy and proof ordinarily required of scientific findings. In the absence of such testing, they fear that irrationality will spread. "There is no guarantee that a society...infected by unreason will be resistant to even the most...dangerous ideological sects," said Dr. Paul Kurtz, professor of philosophy at the University of Buffalo.[3] Nevertheless, with television networks[4] and mass circulation magazines devoting attention to so-called pseudoscientific cults, their influence is on the rise.

Social critic Theodore Roszak traces today's "paranormal boom" to the counterculture movement of the late 1960s, whose adherents openly rejected objective analysis as a means of judging the truth. For a whole class of "bright, widely read, well-educated people" the style was to "accept and endorse all things occultly marvelous." Increasingly, "skepticism [became] a dead language, intellectual caution an outdated fashion...."[5] In the intervening decade, Roszak contends, the public's "powers of amazement" have blurred the distinction between fact and fiction to a point where "the impossible" is now easier to accept than to question.

Renewed Argument Over UFO Sightings

Few scientific arguments in recent times have aroused more public passion than the debate over unidentified flying objects (UFOs). The UFO debate has a relatively long history, but in recent years it has taken on the sort of emotionalism that surrounded the concept of biological evolution a century ago. Since a sighting was first recorded in 1947,[6] data about UFO encounters have run the gamut from the bizarre to the glorious. And while many reports have been patently absurd, some have left a good number of people puzzled by the possibility they could be true. Reports of three sightings have commanded unusual attention:

In May 1952, in clear skies 35,000 feet above Arizona, a U.S. Air Force B-36 bomber encountered two disc-like objects that, in the

---

[3] Quoted in *Time*, Dec. 12, 1977.
[4] "The $6 Million Man," "Wonder Woman," "The Bionic Woman," "The Man from Atlantis," "Star Trek," "Lost in Space," "Space 1999" and "In Search of..." are all popular television shows that deal with paranormal subjects.
[5] Theodore Roszak, *Unfinished Animal: The Aquarian Frontier and the Evolution of Consciousness* (1975), p. 2.
[6] On June 24, 1947, a private airplane pilot, Kenneth Arnold, reported seeing a formation of nine flying objects near Mt. Rainier in Washington. Later, Arnold told reporters the objects resembled "flying saucers."

JIMMY CARTER'S 1973 UFO REPORT

During the autumn of 1973 hundreds of people throughout the United States reported UFOs to NICAP. Among those ~~~~~~~~~~~~~~~~~~~~~~~~~~~~~~ing the recent ~~~~~~~~~~~~~~~~~~~~~~~~~~~~~~ Car~~~~~~~~~~~~~~~~~~~~~~~~ by the accurate~~~~~~~~~~~~~Jimmy Ca~~~~~~ely about a person~~~~~~~~~~~~ its members should have the complete report as it was submitted. President Carter's handwritten report has been typed for clarity.

NATIONAL INVESTIGATIONS COMMITTEE ON AERIAL PHENOMENA (NICAP)®
3535 University Blvd. West
301-949-1267
Kensington, Maryland 20795

REPORT ON UNIDENTIFIED FLYING OBJECT(S)

This form includes questions asked by the United States Air Force and by other Armed Forces' investigating agencies, and additional questions to which answers are needed for full evaluation by NICAP.

After all the information has been fully studied, the conclusion of our Evaluation Panel will be published by NICAP in its regularly issued magazine or in another publication. Please try to answer as many questions as possible. Should you need additional room, please use another sheet of paper. Please print or typewrite. Your assistance is of great value and is genuinely appreciated. Thank you.

1. Name **Jimmy Carter**   Place of Employment

   Address **State Capitol Atlanta**   Occupation **Governor**
   Date of birth
   Education **Graduate**
   Special Training **Nuclear Physics**
   Telephone **(404) 656-1776**   Military Service **U.S. Navy**

2. Date of Observation **October 1969**   Time   AM   PM **7:15**   Time Zone **EST**

3. Locality of Observation **Leary, Georgia**

4. How long did you see the object?_____ Hours **10-12** Minutes_____ Seconds

5. Please describe weather conditions and the type of sky; i.e., bright daylight, nighttime, dusk, etc. **Shortly after dark.**

6. Position of the Sun or Moon in relation to the object and to you. **Not in sight.**

7. If seen at night, twilight, or dawn, were the stars or moon visible? **Stars.**

8. Were there more than one object? **No.**   If so, please tell how many, and draw a sketch of what you saw, indicating direction of movement, if any.

words of a crewman, "seemed to shimmer and dance" beside the plane. A dozen crew members watched as the shapes maneuvered around the midsection of the fuselage for five minutes. After landing, the airmen were questioned by military intelligence officials. Each told the same story.

In October 1974, Air Force Capt. Larry Coyne was flying a helicopter with three crew members aboard at 1,500 feet over central Ohio. Suddenly, Coyne reported, a red light began approaching the chopper "at a great speed on a collision course." When the light came within 500 feet of the helicopter, the crew could make out a metalic gray, cigar-shaped craft. It hovered for several minutes, beaming its bright light into the cockpit, then turned and sped off.

In October 1969, 11 men attending a Lions Club dinner in Leary, Ga., saw a "bluish...then reddish" shape in the sky. It "came close [and] then moved away" before finally disappearing.

The first two sightings are classified among the 20 per cent or so of all UFO reports that cannot be explained by known causes. The third, which turned out to be a sighting of the planet Venus, is noteworthy for another reason. It was made by President Carter.[7]

---

[7] In 1973, Carter filed two formal reports of a UFO sighting that had occurred four years earlier; one was sent to the International UFO Bureau in Edmond, Okla., and the other to the National Investigation Committee on Aerial Phenomena in Kensington, Md.

Occurrences Often Mistaken for UFOs

Ball Lightning. A mass of moving light about the size of a coconut. It gives off a sulphurous odor and frequently explodes in midair. Scientists theorize ball lightning is formed by electrically ionized gases.

Bolides. Colorful meteors that leave a luminous trail as they fall through the atmosphere. Some bolides burn brightly enough to be seen in daylight. Visible over large areas, they usually give the impression of close proximity to widely scattered observers.

Swamp Gas. Glowing gases rising up from marshes that sometimes appear as globes of light. Swamp gas is also called "will o' the wisp" or "witch's light."

Moon Dogs and Sun Dogs. Light reflected from atmospheric ice crystals. As seen from earth, these reflections resemble strange shapes in the sky.

Flights of Birds. Night-flying birds reflecting ground light have often been mistakenly reported as UFOs.

Undoubtedly encouraged by news of Carter's experience as well as his campaign pledge to look into the UFO question,[8] people have deluged the White House with letters requesting that the government "do something" about the possibility that UFOs exist.[9] One course of action, now considered unlikely, is to reopen a formal investigation of the matter. In 1968, the U.S. Air Force released the results of its 22-year study of UFOs, a project known as Operation Blue Book and directed by the late Edward U. Condon, former head of the National Bureau of Standards. The report, "Scientific Study of Unidentified Flying Objects," concluded there was no scientific evidence to support belief in extraterrestrial encounters. But many believers were not dissuaded. They charged that the investigators purposely overlooked material that tended to confirm the existence of flying saucers.

Dr. Frank Press, the President's science adviser, asked the National Aeronautics and Space Administration (NASA) to handle the task of answering a mounting number of inquiries about UFOs. But in December, the agency turned down the request, saying it would probably be "unproductive." *Science* magazine reported NASA officials feared "that reopening the question of the genuineness of visitors from outer space would legitimize a subject most established scientists believe to be phony and a waste of time."[10]

---

[8] Carter was quoted by the *National Enquirer* as having said, "If I become President, I'll make every piece of information that this country has about UFO sightings available to the public. I am convinced UFOs exist because I have seen one." A White House press assistant said she vaguely recalled Carter making the pledge but could not pinpoint the time or place.
[9] See "UFOs Just Will Not Go Away," *Science,* Dec. 16, 1977, p. 1128.
[10] *Ibid.,* p. 1128.

The fact that the scientific community has refused to consider the UFO question worthy of study, however, has not dampened the spirits of private investigators, many of whom interpret this reaction as a cover-up. Claiming that the Air Force study was biased and unscientific, some UFO organizations have called for the formation of a national panel of independent experts to examine the matter of aerial mysteries until all doubts have been put to rest. So far, the government's response has not been favorable.

It would be impossible "to mount a research effort without a better starting point," NASA Administrator Robert Frosch said.[11] NASA makes a distinction, though, between investigating visits from outer space and probing outer space itself for signs of life. Under a newly established Search for Extraterrestrial Intelligence (SETI) program, the agency is expected to request as much as $1 million in the 1979 federal budget for equipment to send radio signals into "deep space" in the hope of hearing a response.

'Out of Body' Experience; Psychic Force

The fascination of many Americans with Eastern religions[12] has helped to create a climate receptive to beliefs and occurrences that fall outside the realm of accepted knowledge. In fact, many of today's psychic practices employ various spiritual techniques to achieve their ultimate goal of tapping the hidden reservoirs of the mind. The "out of body" phenomenon, also known as astral projection, combines elements common to both Eastern and Western spiritual thought. A form of self-hypnosis is said to produce the sensation of being alive and conscious while outside of one's physical body. Some of those who believe they have perfected the ability to transport themselves in this manner also feel they may have discovered the key to the mystery of life after death.

Astral projection has attracted the interest of researchers at the University of Virginia and SRI International (formerly Stanford Research Institute). Dr. Elisabeth Kubler-Ross, a psychiatrist who reports having experienced astral projection, compares it to accounts she has received from a number of patients who had been pronounced clinically dead but survived. Kubler-Ross found that many such patients frequently told of having experienced the sensation of "floating out of their physical bodies" and feeling "a great sense of peace and wholeness." Most of these persons, she said, "were aware of another person who helped them in their transition to another plane of ex-

---

[11] Quoted in *The Washington Post*, Dec. 28, 1977.
[12] See "Eastern Religions and Western Man," *E.R.R.*, 1969 Vol. I, pp. 431-450.

istence. Most were greeted by loved ones who had died before them, or a religious figure."[13]

Some skeptics contend that out-of-body experiences are a type of dreaming conditioned by previously held religious beliefs. Kubler-Ross disagrees. Even subjects with no prior religious convictions, she said, reported the experience gave them a deep sense of "spiritual rebirth." Dr. Kubler-Ross said she was at first suspicious of the phenomenon but after studying it became convinced "there is life after death beyond the shadow of a doubt."[14]

Another theory that has gained wide popularity in the past several years is the "biorhythm" or "biocurve" interpretation of human performance. Derived from the experiments of 19th century German psychologist Wilhelm Fleiss, the concept holds that the rise and fall of physical, intellectual and emotional energy levels proceed in cycles of 28, 33 and 38 days. Industrial productivity studies and the performance of athletes on so-called "up" and "down" days are offered as evidence to support success claims made by promoters of biorhythmic charts. But scientists question the basic assumption of the theory, which proposes that all human behavior adheres to standard patterns regardless of the age, sex or life events of the subjects.

Enjoying even less favor among scientists and physicians is the practice of psychic healing—sometimes called psychic surgery. Its practitioners reject the usual tools of medicine and rely on psychic forces to diagnose and allegedly cure illnesses. These practitioners claim they can perform many types of "operations." However, Dr. William A. Nolen, an American surgeon, who analyzed the methods of psychic surgery in his book *Healing* (1975), suggested that they rely almost entirely on slight-of-hand deception. He contends, for instance, that animal organs are often passed off as excised human tissue.

Opposition From a New Science Group

A group of 43 scientists, writers and educators, concerned about what they considered the growing acceptance of mysterious beliefs, in 1976 organized the Committee for the Scientific Investigation of Claims of the Paranormal. Made up of a number of noted figures, including astronomer Carl Sagan, psychologist B. F. Skinner and author Issac Asimov, the committee twice yearly publishes a journal, *The Zetetic* (Greek for "skeptic"), that features critical analyses of various paranormal

---

[13] Elisabeth Kubler-Ross, "Foreword" in *Life After Life* (1975), a book by Dr. Raymond A. Moody, who reported a collection of similar cases. Kubler-Ross is perhaps best known for her book *On Death and Dying*, a pioneering study of the experiences of dying people.
[14] Quoted by Ann Neitzke in "The Miracle of Kubler-Ross," *Human Behavior*, September 1977, p. 25.

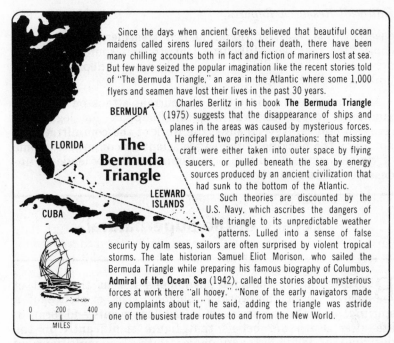

Since the days when ancient Greeks believed that beautiful ocean maidens called sirens lured sailors to their death, there have been many chilling accounts both in fact and fiction of mariners lost at sea. But few have seized the popular imagination like the recent stories told of "The Bermuda Triangle," an area in the Atlantic where some 1,000 flyers and seamen have lost their lives in the past 30 years.

Charles Berlitz in his book **The Bermuda Triangle** (1975) suggests that the disappearance of ships and planes in the areas was caused by mysterious forces. He offered two principal explanations: that missing craft were either taken into outer space by flying saucers, or pulled beneath the sea by energy sources produced by an ancient civilization that had sunk to the bottom of the Atlantic.

Such theories are discounted by the U.S. Navy, which ascribes the dangers of the triangle to its unpredictable weather patterns. Lulled into a sense of false security by calm seas, sailors are often surprised by violent tropical storms. The late historian Samuel Eliot Morison, who sailed the Bermuda Triangle while preparing his famous biography of Columbus, **Admiral of the Ocean Sea** (1942), called the stories about mysterious forces at work there "all hooey." "None of the early navigators made any complaints about it," he said, adding the triangle was astride one of the busiest trade routes to and from the New World.

topics. In a recent issue, for example, a member wrote that the fork-bending feats of psychic Uri Geller could be performed by almost any stage magician.[15]

A current target of committee criticism is an NBC-TV show "Exploring the Unknown," which was broadcast last fall and dealt with such subjects as levitation and contact with the dead. The committee formally complained to the Federal Communications Commission that the show presented "a totally biased" point of view. A committee representative, Dr. Paul Kurtz, said that NBC's "documentary" treatment of the Bermuda Triangle *(see box above)*, UFOs and other similar subjects "constitutes in scientific terms a scandal."

A spokesman for the network, George Hoover, defended the show, saying it was intended to be seen as entertainment, not documented fact. "The only thing we call a documentary is a program that's produced by NBC News." And what is not identified as news by the network should not necessarily be assumed to be true, Hoover added.

In answer to the contention that public interest in psychic phenomena is basically harmless, committee members have argued that blind acceptance of the paranormal may expose certain people to serious hazards. They say the promise of miracle cures has persuaded many persons suffering from illness or dis-

---

[15] See *The Zetetic*, spring-summer 1977. Uri Geller is an Israeli who claims to be able to perform various psychic feats. Geller's ESP abilities have been tested at SRI International (formerly Stanford Research Institute) with positive results. But critics have challenged those experiments on the grounds that they were not properly conducted.

ease not to accept traditional medical treatment. According to committee member Milbourne Christopher, some 200 people are known to have killed themselves because of depression over unfavorable horoscopes.[16]

Many scientists believe that the current attack on scientific knowledge could undermine public support for legitimate research efforts. Nevertheless, the work of the committee does not enjoy the backing of the entire scientific community, some of whose members believe formal opposition only lends an undeserved dignity to the new cults.

Legacy of the Supernatural

B ELIEF in supernatural phenomena lies at the heart of nearly all of the world's religions. The existence of a spiritual soul, the occurrence of miraculous events and forms of life after death are beliefs that figure significantly in the doctrines of virtually every major faith. For multitudes of Christians, the transformation of bread and wine into the body and blood of Christ is a central theological tenet. Christian dogma also holds it possible for human beings to have spiritual experiences, such as those recounted in the lives of the saints, that surpass common understanding.

Many cultural anthropologists suggest that belief in the presence of supernatural forces is so general throughout the world that it must be considered an instinctive aspect of human nature. Author Colin Wilson wrote: "[P]rimitive man believed the world was full of unseen forces: the *orenda* (spiritual force) of the American Indians, the *huaca* of the ancient Peruvians. The Age of Reason said that these forces had only...existed in man's imagination; reason [alone] could show man the truth about the universe."[17]

Yet there is an "immense world" of intuition that stretches far beyond the rational, Wilson added. In Western literature, perhaps Goethe's epic drama *Faust* best portrays the conflict between man's desire to call forth the occult powers of the mind and his opposing urge to heed the restraints of logic and reason.

In every age, there are those who have attempted to come to terms with the mysteries of clairvoyance, telepathy and divination. However, efforts to explore these forces have often

---

[16] Reported in *The New York Times*, Oct. 10, 1977.
[17] Colin Wilson, *The Occult* (1971), p. 21.

run counter to established religious norms. Until modern times, it was rarely argued that a competing body of knowledge invoking the supernatural did not work, but only that it was evil.

Injunctions against rival spiritual practices are found throughout the Old Testament. In the Book of Deuteronomy, Moses warns the Israelites that witches, wizards, and necromancers are "an abomination unto God." Similarly, the teachings of other religions caution followers against false prophets and conjurors of evil spirits. Regardless, varieties of occult beliefs have survived down through the centuries. And they continue to thrive even in the world's most technically advanced nations—vivid, sometimes disturbing, reminders of man's ancient hopes and fears.

Folktales About Werewolves and Witches

Long before the Christian era, there were legends about men and women whose contact with evil forces gave them dark supernatural powers over death. It was author Bram Stoker who was mainly responsible for reviving one such legend—the myth of the werewolf—in his novel *Dracula,* which appeared in 1897. In Western lore, tales of vampires first appeared in ancient Greece and centered on the belief that if a dog or cat jumped over a corpse, the body would be recalled to life and compelled to seek out human victims in order to satisfy its need for blood.

That belief persisted for centuries in areas of Eastern Europe where the werewolf played a prominent part in local folklore. People thought to have been killed by the bite of a werewolf were buried with a consecrated host on their chests to insure that they would not become one of the creatures. Rumors of vampirism were widespread in the mountainous regions of Germany and Hungary during the 18th and early 19th centuries. Documents recount the opening of graves of suspected werewolves and the discovery of bodies contorted in postures of horrible agony. This was thought to be a sure sign of the "vampire's curse." Modern skeptics, however, reject such notions, suggesting instead that the contortions may have been the consequence of premature burial, not uncommon in the days before the embalming of dead bodies was widely practiced.

Popular belief in the evil powers of witches can be traced to the worship of primitive fertility gods, once considered the source of both good and bad fortune. From its early times, the Church has proscribed witchcraft, which was frequently associated with pagan sexual rites. Throughout the Middle Ages, heretics and other rebels against accepted religious custom were often put to death as witches, as were political plotters during the Protestant Reformation.[18]

[18] See "The Occult vs. the Churches," *E.R.R.,* 1970 Vol. I, pp. 299-318.

Men who practiced the so-called black arts are usually referred to as wizards or warlocks. The term "witch" is applied almost exclusively to women. It was commonly thought that witches derived their evil powers from sexual relations with demons. Oddly enough, however, England saw "few accusations that witches had actually been in direct contact with the Devil, as was [the case] in some parts of Europe.... Most English witches were accused of damaging their neighbor's property. So complex were laws on the subject it was nearly impossible for the accused person to escape conviction."[19]

English laws against the practice of witchcraft date from the reign of King Canute (1000-1036). Until they began to disappear in the mid-18th century, both civil and ecclesiastical witch trials were commonplace. Between 1558 and 1736, in London alone, some 200 persons were convicted of using sorcery and black magic. The most notorious American witch trials occurred in Salem, Mass., in 1692. Nineteen persons were hanged and one was pressed to death following accusations that they had communicated with evil spirits.

Alchemy: the Practical Use of Magic

Ancient priests and magicians tended to regard the world as a vast allegory in which visible things reflected the presence of invisible things; real objects and events were thought to be connected in some obscure way to corresponding spiritual forms. It was believed, and of course many people still believe, that extraordinary powers are available to those who master this sytem of correspondences. The incantations and arcane formulas the alchemists used to enlist the forces of the supernatural can be seen as a manifestation of this idea.

The practice of alchemy combined both technical and mystical knowledge in an attempt to transform base metals such as lead and copper into silver and gold. In much of their work, the alchemists represent a significant link between the occult beliefs of magic and early forms of practical science. In fact, for many centuries, the history of alchemy and the history of chemistry were identical.

The origins of alchemy remain a mystery. There is speculation that the art arose in Egypt and was eventually carried to Greece where it was practiced by artisans and scholars who attached themselves to wealthy patrons. It was the Greek philosopher Aristotle whose ideas on the composition of matter supplied the theoretical basis for alchemy. Aristotle taught that all matter was composed of varying amounts of four elements:

---

[19] Derek and Julia Parker, *The Immortals: The Mysterious World of Gods, Gobblins, Fairies, Leprechauns, Vampires, and Deities* (1975), p. 173.

earth, water, air and fire. For example, solid objects were composed primarily of earth; liquids primarily of water; the spirit was made up of air; fire was seen as the catalyst of change. Conceivably, any material substance could be converted into any other if the correct arrangement of elements could be achieved.

Not surprisingly, the first Greek alchemists were metal workers, men familiar with the process of melting and combining substances. To Aristotle's assumptions about matter, they added the astrological notion that under the proper heavenly influences man could effect changes in less valuable metals by burning away their inferior qualities and remaking them in purer forms. Although the alchemists engaged in a rudimentary type of chemistry, they were guided by principles of magic and mystical knowledge, which they sought to apply in practical ways.

From its beginning, the Catholic Church regarded alchemy as the Devil's work. To be sure, many alchemists were charlatans who preyed upon the greedy instincts of their admirers. But quite frequently there were also men who enjoyed considerable scholarly respect. Alchemy, "once by the judgment of all old philosophers..., held the highest place of honour," wrote a 15th century Florentine historian, Cornelius Agrippa. However, from the "first days of the Church it has been forbidden most unjustly: for it is holy learning."[20] Alchemy had long been looked upon as a sort of underground art, but by the end of the 16th century a resurgence of militant religious feelings during the Reformation marked its demise in most parts of Europe.

Astrology's Vision of an Ordered Cosmos

Magic and astrology, while they are sometimes confused in practice, differ in one essential respect, according to C. S. Lewis, the English man of letters. "The magician asserts human omnipotence; the astrologer, human impotence. The thoroughgoing astrologer is a determinist. He shatters the illusions and exciting hopes of the magician. Those temperaments that are attracked by modern forms of determinism in our own day would have been attracted by ancient astrological determinism."[21]

Astrology may be the oldest of today's occult "sciences." Its roots go back 5,000 years to Babylonia where stargazers began noting specific conjunctions of heavenly bodies and favorable or unfavorable earthly occurrences. Using the zodiac, a device for plotting the imagined relationship of stars to the Earth, the

---

[20] Cornelius Agrippa, *De Occulta Philosophia*, quoted in C. S. Lewis's *English Literature in the Sixteenth Century* (1954), p. 9.

[21] C. S. Lewis, *op. cit.*, p. 6.

Babylonians identified 12 stellar groups transversed by the sun and planets. Each separate constellation was named for an animal or human figure which its outline seemed to resemble.

The Three Wise Men who visited the infant Jesus, according to the New Testament stories, were though to be astrologers drawn to Bethlehem in search of a momentous event foretold by what they took to be a new star. Centuries later, however, the Catholic Church condemned the practice of astrology. St. Augustine insisted in *The City of God* that if astrological predictions happened to come true it was due either to chance or the intervention of demons. In 1586, Pope Sixtus V issued a papal decree against astrology. Martin Luther and other Protestant churchmen of the age took similar stands.

Astrology gradually fell into disfavor in the late 17th century as the Age of Reason began. Advancements in the science of astronomy revealed that the stars were not orderly and timeless as astrologers thought, but instead were bodies in constant flux. Dismissed by most reputable scholars, astrological writings nevertheless continued to have a strong influence on many artists and poets, including some as late as William Blake and William Butler Yeats.

Today, the appeal of astrology is principally that of a "specious psychology," wrote columnist George F. Will. In an age of uncertainty and rapid social change, astrology "enables people to classify themselves.... [It] bestows on its believers a sense of being not quite completely adrift on turbulent seas. Many people find it soothing to believe that they are under the predictable sway...of common forces," Will concluded.[22]

Debate on Psychic Concepts

FROM THE GREEK oracle at Delphi to today's tabloid soothsayers, prophets and seers have played a major role in the history of occultism. By far, the most popular of the current predictors in America is the self-professed seeress Jeane Dixon. Although she has been making her predictions public since the early 1950s, Dixon achieved national prominence after the assassination of President Kennedy in 1963, which she claimed she foretold seven years before. This prediction was published in *Parade* magazine on May 13, 1956, and read as follows: "As for the 1960 election...it will be dominated by labor and won by a Democrat. But he will be assassinated or die in office, though not necessarily in his first term." It should be noted

---

[22] George F. Will, writing in *The Washington Post*, Sept. 15, 1975.

Psychic Practices and Beliefs

Clairvoyance. Extrasensory perception of objects and events, as distinguished from thoughts and mental states.

Extrasensory Perception (ESP). Ability to know something that does not seem to involve any physical senses or rational inference.

Extraterrestrial Descent of Man. The belief that ancient astronauts landed on earth and peopled the legendary island of Atlantis.

Kirlian Photography. Technique of photographing so-called energy fields and emotional auras said to emanate from human subjects.

Psychokinesis. Direct mental influence on a physical object.

Pyramid Power. Belief that the Great Pyramids of Egypt and pyramid-shaped objects are a source of energy that can be used to increase emotional, intellectual and physical powers.

Rolfing. Practice developed by Dr. Ida Rolf, a psychologist, that involves the manipulation of muscles to put the body into natural alignment with the earth's gravity.

Silva Mind Control. A form of self-hypnosis through which users claim they can diagnose and cure illnesses by psychic forces.

that Dixon predicted in 1960 that Kennedy would fail to win the presidency.

Her views and forecasts have received wide circulation through her various books and on a regular basis in the weekly paper the *National Enquirer.* Dixon gives lectures across the country and in some cities the telephone company provides her special daily "Horoscopes by Phone" as a public service. Her Washington office reports that 300 newspapers subscribe to her syndicated horoscope column.

Dixon claims that her predictions are divinely inspired by God. When she is wrong, she has said, it is because she failed to read God's signs correctly. Unquestionably, she has attracted a large following. Yet for all the notice given to her famous thoughts on the future, particularly those made at the beginning of each new year, little attention is paid to how they come out. "No one has really assembled Mrs. Dixon's misses," [23] complained critic Hugh Tyler. Dixon's method, her detractors argue, is to make so many predictions that a few are bound to be right. Those, they say, are the ones that are publicized, while the hundreds that go wrong are forgotten. The following are some of her prognostications for 1976 that went awry:

Ex-President Nixon would make a political comeback.

---

[23] Hugh Tyler, "The Unsinkable Jeane Dixon," *The Humanist*, May-June 1977, p. 6.

TV's bionic man, actor Lee Majors, would lead the way in promoting bionic motors for crippled people.

Jacqueline Onassis would remarry.

Hijackers would take over a subway and hold its passengers for ransom.

In March 1976, Dixon said there would be an assassination attempt on then-President Ford before the year was over. There were two assassination attempts on Ford in 1975, but none in 1976. She also predicted there would be an attempt on Ronald Reagan's life and that Reagan would win the 1976 Republican nomination over Ford. Neither event occurred.

Like other seers, Dixon deals mostly in predictions of disasters and calamities. "One thing that makes me suspicious of prophets is that they all use the same system," said Daniel St. Albin Greene, author of a number of articles critical of the claims of popular mystics and seers. "They're always predicting sickness or assassinations or global problems. A person like Jeane Dixon is constantly telling people that something dire is going to happen. Most of these things do not happen. But occasionally they do, and then [some people] swear to God that Jeane Dixon prophesied it."[24]

For believers, Jeane Dixon is the unimpeachable voice of the future. As for her own personal future, Dixon has predicted she will spend her next life in outer space. In 1976, she said she envisioned herself in a rocket ship among "feminine creatures in metallic one-piece garments" who communicated "entirely by telepathy."[25]

Scientology's View of Psychosomatic Ills

Americans seem to have an insatiable passion for self-improvement. The search has always been for sure ways of achieving success or "winning friends and influencing people." Increasingly, methods offered to help reach the goal of a more rewarding life are being advertised as "scientific." In a field rife with competition, the word evokes a special kind of legitimacy. Roy Wallis of the London School of Economics wrote:

> During the early nineteenth century, as medical practice became more competent to deal with physical illness, expectations of health and well-being...centered on the psychological domain and the difficulties of interpersonal relations. Movements, like Christian Science..., that have claimed efficacy in handling physical illness, lost ground, while others arose offering psychological well-being, release of mental and emotional tension, cures for psychosomatic and neurotic illnesses, techniques for releasing hidden inner abilities....

---

[24] Quoted in *The Humanist, op. cit.,* p. 9.
[25] Writing in the *National Enquirer,* Nov. 9, 1976.

In such areas science has as yet made little concrete progress. But the prestige of science has become such as to require that almost every new movement entering this field claim scientific...authority, if by no other means than that of incorporating the word 'science' in its title.[26]

One such practice that blends the tools of confidence-building with the trappings of psychomedicine is scientology. Founded by L. Ron Hubbard, a Nebraska-born salesman, who now lives in England, scientology is regarded by its adherents as a religion, although the Internal Revenue Service thinks otherwise. Since it was established nearly 25 years ago, the scientology movement has been involved in battles with several government agencies over whether it should be granted tax-exempt status as a non-profit church.

"Man does not live by reason alone; and science is often viewed by critics as cold and rational. People hunger for something more."

Paul Kurtz
The Humanist, July 1976

The main principles of scientology are based on a system of psychic practices called dianetics, developed by Hubbard in the late 1940s. He theorized that the mind is divided into two parts: the "analytical" and the "reactive." The analytical mind is described as an efficient, rational instrument. But Hubbard suggested it rarely works at full capacity because of interference by its counterpart, the reactive mind—a "memory bank" of painful past experiences known as "engrams." The objective of scientology is the removal of these "engrams." The process involves long and costly sessions with "auditors," who draw them out and erase them one by one until the subject is completely cleared. At this point, it is claimed IQ increases and the subject becomes immune to diseases of the mind and body.

There is much about scientology that is unknown, since Hubbard has kept the organizational aspect of the movement secret. But most critics generally see the practice of dianetics as a form of amateur psychology. Many physicians also question claims that it cures arthritis, tuberculosis, ulcers and dozens of other ailments that Hubbard defines as psychosomatic.

---

[26] Roy Wallis, "Poor Man's Psychoanalysis? Observations on Dianetics," *The Zetetic,* fall-winter 1976, p. 19.

Recently, the federal government has begun to investigate scientology for reasons not directly related to its philosophy. FBI raids last summer in several cities turned up alleged evidence that members of the movement had been stealing secret government documents and "bugging" and breaking into government offices. Federal officials speculate that there may be a connection between these alleged activities and the scientology movement's long fight with the IRS over the question of taxes.

Assessing the Effects of Unproven Claims

Thomas Macaulay, the 19th-century British historian, believed that as civilization progressed, the need for poetry and imagination declined; the more men understood about the world, the less mystery they saw in it. Macaulay's dictum was a particularly apt description of the 19th century when new scientific discoveries replaced dozens of "old beliefs" with proven facts. Science and religion, reason and imagination have always been at odds. Philosopher Paul Kurtz has written: "Theologians have incessantly argued that there are 'limits' to scientific inquiry and that it cannot penetrate the 'transcendental realm'; and poets have decried deductive logic and the experimental method, which they claim denude experience of its sensuous qualities."[27]

Confidence in science, however, has been eroded in recent years, partly perhaps by an increasing aversion to technological culture and partly by what observers see as a trend toward anti-intellectualism, a general movement in society away from the values of critical judgment and skepticism. Logic has fallen into disrepute, a turn of events, some scientists say, was encouraged by the easy credence given to "subjective truth." In many quarters, the scientific approach has come to be seen as too narrow and limiting.

It is more convenient, for example, to accept the mysterious disappearance of ships and planes in the Atlantic Bermuda Triangle than to examine the "basic facts" of each case, wrote author Larry Kusche. "Ignorance, in fact, is a major technique in writing about the mystery of the Bermuda Triangle and other subjects in the so-called paranormal as well. Some critics refer to [these subjects] as science fiction. But that is an unfair description. Unfair, that is, to science fiction. The topics might more properly be called fictional science."[28]

---

[27] Paul Kurtz, "The Scientific Attitude vs. Antiscience and Pseudoscience," *The Humanist*, July-August 1976, p. 27.

[28] Larry Kusche, "Critical Reading, Careful Writing, and the Bermuda Triangle," *The Zetetic*, fall-winter 1978, p. 39. Kusche is author of *The Bermuda Triangle Mystery—Solved* (1975).

The historical role of science has been to question and test assumptions. Today, though, the scientific establishment finds itself in the position of having to defend its own methods. "To deviate from objective thinking is to be out of touch with cognitive reality," Paul Kurtz has written. "The paradox is that so many people are willing to abandon the use of their practical intelligence when they enter fields of religion or ethics, or to throw caution to the wind when they flirt with so-called transcendental matters."[29]

While many scientists accuse proponents of paranormal phenomena of charlatanism, scientists are often criticized in turn for being too censorial and intolerant of new discoveries. Somewhere between these two opposing factions are those who warn there is as great a danger in extreme skepticism as there is in irrational belief.

"In the past few centuries," Colin Wilson wrote, "science has made us aware that the universe is stranger and more interesting than our ancestors realized. It is an amusing thought that it may turn out stranger and more interesting than even the scientists are willing to admit."[30] Most scientists recognize that possibility. But at the same time, they maintain that the best way to unlock the secrets of the paranormal, whatever they may be, is with the same tools of logic and reason practical knowledge has traditionally depended on.

---

[29] Paul Kurtz, *op. cit.*, p. 30.
[30] Colin Wilson, *The Occult* (1971), p. 33.

Selected Bibliography

Books

Condon, Edward U., et al., *Scientific Study of Unidentified Flying Objects,* Bantam, 1969.

Eiseley, Loren, *The Immense Journey,* Random House, 1957.

Hansel, C.E.M., *ESP: A Scientific Evaluation,* Scribner, 1966.

Jacobs, David, *The UFO Controversy in America,* University of Indiana Press, 1975.

Koestler, Arthur, *The Roots of Coincidence,* Vintage, 1972.

Moody, Raymond A., *Life After Life,* Bantam, 1975.

Nolen, William A., *Healing: A Doctor in Search of a Miracle,* Fawcett, 1975.

Panati, Charles, ed., *The Geller Papers: Scientific Observations on the Paranormal Powers of Uri Geller,* Houghton Mifflin, 1976.

Saunders, David R., and R. Roger Harkins, *UFO's? Yes! Where the Condon Committee Went Wrong,* Signet, 1968.

Shklovski, I.S., and Carl Sagan, *Intelligent Life in the Universe,* Dell, 1966.

Wilhelm, John, *The Search for Superman,* Pocket Books, 1976.

Articles

Batson, Daniel, "Moon Madness: Creed or Greed," *Monitor* (a publication of American Psychological Association), June 1976.

Gumpert, Martin, "The Dianetics Craze," *New Republic,* Aug. 14, 1950.

Jacobson, Laurie, "Feedback on Biofeedback," *Human Behavior,* July 1974.

Kurtz, Paul, "The Scientific Attitude vs. Antiscience and Pseudoscience," *The Humanist,* July-August 1976.

Kusche, Larry, "Critical Reading, Careful Writing, and the Bermuda Triangle," *The Zetetic,* fall-winter 1977.

Lawrence, Jerome E., "Astrology: Magic or Science?" *The Humanist,* September-October 1975.

Nietzke, Ann, "The Miracle of Kubler-Ross," *Human Behavior,* September 1977.

Omohundro, John T., "Von Daniken's Chariots: A Primer in the Art of Cooked Science," *The Zetetic,* fall-winter 1976.

Tyler, Hugh, "The Unsinkable Jeane Dixon," *The Humanist,* May-June 1977.

"Why Scientists Take Psychic Research Seriously," *Business Week,* Jan. 26, 1974.

Reports and Studies

Editorial Research Reports, "The Occult vs. The Churches," 1970 Vol. I, p. 299; "Psychomedicine," 1974 Vol. II, p. 987; "Approaches to Death," 1971 Vol. I, p. 287.

U.S. Air Force, "Scientific Study of Unidentified Flying Objects," 1968.

U.S. House of Representatives Committee on Science and Astronautics, "Symposium on Unidentified Flying Objects," 1969.

Computer Crime

by

Marc Leepson

**Jan. 6
1978**

Editor's Note: The Senate Judiciary Committee held hearings, June 21-22, 1978, on a bill introduced by Sen. Abraham Ribicoff, discussed on p. 48, to authorize long prison terms and heavy fines for nearly all unauthorized uses of government-owned computers and privately owned computers involved in interstate commerce. The legislation currently is pending in the committee.

COMPUTER CRIME

T HE ANNALS of crime are recording a new and growing type of criminal activity: crimes involving computers. Fraud, embezzlement, blackmail and other crimes committed by the manipulation or misuse of computers cost Americans more than $100-million a year, according to the U.S. Chamber of Commerce.[1] "Today business and government are more vulnerable to white-collar crime through use of computers than they were ever before or probably ever will be in the future," according to Donn B. Parker, senior management systems consultant at SRI International (formerly Stanford Research Institute) in Menlo Park, Calif. Parker, an expert on computer fraud, said a basic reason for this vulnerability is "the lack of progress in recognizing the threat and taking protective action in a period of rapid transition from manual, paper-based business activities" to fully computerized systems.[2]

There are other reasons why computer-related crime is on the increase. For one thing, the number of computers and persons who work with them is rising steadily. International Data Corp., a publishing and market research consulting firm with headquarters in Waltham, Mass., reported that 86,314 general-purpose computers were installed in American businesses as of Jan. 1, 1977—the latest date for which figures are available. The company also reported that 176,315 minicomputers—small, relatively inexpensive units—are in use. In addition, the U.S. government uses some 10,000 computers.

Computers touch the daily lives of nearly all Americans. They are used in nearly all business and governmental functions that are particularly susceptible to monetary theft. They are used by banks, public utilities, consumer credit companies and by financial offices in large corporations and in state, local and federal governments. Along with this increasing use of computers is a parallel rise in the number of persons who work with the machines—operators, programers and technicians. SRI International estimated that 2,230,000 Americans worked directly with computers in 1975. The figure is believed to be substantially higher today.

---

[1] Figure published in the U.S. Chamber of Commerce's 1974 report, "Handbook on White-Collar Crime." Other estimates run as high as $300-million a year.
[2] Donn B. Parker, *Crime by Computer* (1976), p. 298.

In many organizations, management supervisors have only faint knowledge of computer operations. That fact, combined with the near impossibility of checking the extremely complicated computer operating procedures, makes computers infinitely more vulnerable to misuse than the manual paper-based systems they replaced. "It is almost universally conceded in the electronic data processing industry that...it is extremely difficult to detect acts of embezzlement, fraud or thievery in which computers...are used as the principal tools of crime," Thomas Whiteside wrote in a recent *New Yorker* magazine series on computer crime.[3] For example, a bank teller with access to the bank's computer makes transactions that are fundamentally different from those that are written on paper. The ease of access to computers leaves virtually no trails for auditors or other investigators to follow.

Difficulty of Detection and Conviction

August Bequai, a criminal lawyer who served as chairman of a Federal Bar Association subcommittee on white-collar crime, told Editorial Research Reports that the chance of an electronic crime being discovered is only one in a hundred. "And the likelihood of being convicted of a computer crime is one in five hundred and of going to jail one in a thousand. The odds for avoiding a stiff sentence are even more favorable."

Bequai and others have pointed out that the criminal justice system was set up to deal mainly with crimes of violence. But most white-collar criminals, especially computer culprits, are middle-class citizens who typically have no past record of criminal activity. Some 40 statutes are used to prosecute computer crimes but none of these laws was written specifically to cover these crimes. That situation contributes to the difficulties of administering justice to computer criminals.

Sen. Abraham Ribicoff (D Conn.) has introduced a bill in Congress to authorize long prison terms (up to 15 years) and heavy fines (up to $50,000) for nearly all unauthorized uses of government-owned computers and privately owned computers involved in interstate commerce. Ribicoff introduced the bill June 27, 1977, and a similar measure subsequently was introduced in the House of Representatives. No hearings had been held by the Judiciary Committees of either house when Congress ended its 1977 session in December but there was a prospect of action early in the 1978 session, possibly in February. One reason for the slow progress to date is the complex process of integrating the measures into the U.S. criminal code.

---

[3] Thomas Whiteside, "Dead Souls in the Computer," *The New Yorker,* Aug. 29, 1977, p. 34.

Sabotage and Vandalism by Employees

The many types of computer crime fall into two broad categories: (1) vandalism and sabotage, and (2) theft, fraud and embezzlement. Vandals and saboteurs have physically attacked computers on several occasions and for varied reasons. Some attacks have been politically motivated; some have come for personal reasons. A good number of vandalism cases involve disgruntled employees who steal or destroy computer equipment and tapes. Unhappy employees have attacked company computers with shotguns, screwdrivers and gasoline bombs. Malicious workers have deliberately mislabeled, misfiled or erased computer tapes.

A computer operator for Yale Express System in New York City, for example, thought he was being worked too hard and took revenge by destroying billing information he was supposed to enter into the trucking company's computer. He destroyed some $2-million worth of bills. A small business on the West Coast was hit even harder. An unhappy employee programed the company's computer to destroy all accounts receivable six months after he quit his job. This left no record of who owed the company money. It placed a newspaper advertisement asking its customers to pay what they owed. When only a few responded, bankruptcy followed.

Several instances of politically motivated attacks on computers occurred during the Vietnam War. A group of protestors erased computer tapes containing data on napalm production at Dow Chemical Co. offices in Midland, Mich., in 1970. Students attacked Department of Defense research computer centers at the University of Wisconsin and Fresno State College in 1970. The attacks caused millions of dollars in damage at both schools and, in the Wisconsin attack, resulted in the death of a researcher.

August Bequai said recently: "What will prevent a politically motivated group from locking itself into a key computer section of a large multinational or American company and threatening to blow themselves up unless a ransom is paid?" And new ways to apply crimes of sabotage and vandalism to computers may develop. Small, accurately guided weapons, so-called "smart bombs," conceivably could be used to destroy a computer operation or computer tapes from areas many miles away.

It is even possible to commit a murder by computer. Such a crime can take place because computers are capable of controlling life-support systems in hospitals. If a hospital computer guiding a patient's life-support system is tampered with to function incorrectly, the result could be death for the patient. What worries law-enforcement authorities about computer crimes involving vandalism and sabotage is that no special knowledge of computers may be needed. "If technical expertise is lacking," Arthur R. Miller wrote, "a match, or a hammer in the case of a disc or a data cell, will do the same job...in a minute or two."[4]

Electronic Theft, Fraud and Embezzlement

There are many kinds of computer theft, including theft of data, services, property and financial theft. "Time-shared" systems, in which several companies make use of the same computer, offer a particularly tempting target for the theft of services. One of the soft spots in such systems occurs when computerized information moves from the central processing unit *(see p. 70)* through communications links to customers. In addition to the relatively simple task of bugging the transmission line and recording the electronic communications passing over it, the wiretapper might attach his or her own computer terminal to the line and join the group sharing the system's ser-

---

[4] Arthur R. Miller, *The Assault on Privacy* (1971), p. 28.

Reported Cases of Computer Misuse

Year	Vandalism	Information or Property Theft	Financial Fraud or Theft	Unauthorized Use or Sale of Services	Total
1965	—	1	4	3	8
1966	1	—	1	—	2
1967	2	—	—	2	4
1968	2	3	7	1	13
1969	4	6	3	2	15
1970	8	5	10	10	33
1971	6	19	23	6	54
1972	15	18	16	17	66
1973	11	20	26	11	68
1974	7	15	25	12	59
1975	6	7	26	4	43
Total	62	94	141	68	365

SOURCE: SRI International

vices. A bookmaker was caught using a college computer in this manner to run an illegal bet-taking operation.

One of the most publicized cases of computer-abetted property theft occurred in Los Angeles. Jerry Neal Schneider masterminded a scheme which resulted in his conviction for stealing an estimated $1-million worth of telephone equipment from the Pacific Telephone & Telegraph Co. in 1970-1972. Schneider accomplished what author Gerald McKnight called "one of the most amazing robberies in the history of crime"[5] by using stolen computer information to order equipment through the Los Angeles telephone company's own computer. Before he was discovered, Schneider employed 10 persons to gather and sell the pirated equipment.

Confidential business information attracts thieves because it is likely to be very valuable. Large volumes of paper material can be boiled down and stored on small reels of magnetic tape, making it comparatively easy to steal. Thieves stole two reels of Bank of America tape at the Los Angeles International Airport in 1971 and threatened to destroy them if the bank did not pay a large ransom. In January 1976, the head programer of the computer department of Imperial Chemical Industries in Rotterdam, the Netherlands, took all the company's computer tapes that dealt with European operations. He asked for the equivalent of $200,000 in cash for their return.

[5] Gerald McKnight, *Computer Crime* (1973), p. 33.

Computer theft also lends itself to blackmail. Persons with access to computers have retrieved private data and threatened to make the information public. Such potentially damaging information includes poor college performance, erratic employment history, crime conviction or mental hospitalization. In one case, computer-room employees in Manchester, England, threatened to destroy their company's records if it did not give large salary increases.

The most costly, and perhaps the most common, computer crimes involve fraud and embezzlement. There are many cases of clerks with knowledge of their company's computers who have transferred money from customer accounts to their own pockets. But computer fraud does not always involve company clerks or bank tellers or persons working in computer rooms. Some schemes involve perpetrators with no access to computers or computer terminals. One such fraud has taken place in several major cities across the country.

It works like this: The criminal opens a small checking account in a bank and then steals some blank checking deposit slips—the ones that banks provide for depositors who do not have premarked checking deposit forms. The criminal takes the blank forms to an unscrupulous printer who adds the criminal's magnetic-ink checking account number to all the slips. The doctored deposit forms are slipped back into their usual place at the bank. Everyone using them is therefore depositing money into the criminal's checking account. Within days, hundreds of thousands of dollars are sent by the bank's computer into the perpetrator's account. The thief withdraws the money and moves on.

By far the biggest computer-related crime on record involved a nationwide investment firm, the Equity Funding Corporation of America. The Equity Funding scandal first came to public attention in 1973. Computer crime was but one part of a widespread illegal financial fraud masterminded by the highest officers of the now-bankrupt company. Equity's president and 21 other executives were convicted in 1975 of setting up life insurance policies for some 56,000 fictitious persons and selling the policies to other insurance companies. The bogus insurance policies existed only inside Equity Funding's computer. A bankruptcy trustee's report calculated that the total value of the fraudulent policies was $2.1-billion.

Computer fraud is by no means relegated to private enterprise. The federal government, too, has been the victim of computer crime. The General Accounting Office, Congress's investigating arm, reported in 1976 on 69 instances of improper use of computers in the U.S. government. The 69 cases resulted in losses of some $2-million. The report said that the 69 cases "do not represent all the computer crimes involving the federal government since agencies do not customarily differentiate between computer-related and other crimes."[6] The GAO investigation further revealed the distinct possibility that a large number of computer crimes in the federal government have yet to be detected or reported. Computer-related crimes were documented at all levels of the government.

In one case, an Internal Revenue Service programer fed information into an IRS computer to funnel unclaimed tax credits into a relative's account. Another IRS programer used a computer to take checks being held for those whose mailing addresses could not be located and deposit the checks into his own account. Concern about the vulnerability of the federal government's computers is high because the government pays tens of billions of dollars by computer every year. The crimes documented by the GAO ranged from those involving hundreds of dollars to those involving hundreds of thousands of dollars.

Potential for Computer Misuse

R ESEARCH initiated during World War II led to development of the first electronic computer, the Electronic Numerical Integrator and Calculator—commonly known as Eniac—at the University of Pennsylvania in 1946. The Eniac, which took two-and-one-half years to build, solved its first problem, an equation involving atomic physics, in two hours. The machine contained 18,000 vacuum tubes and could carry out 5,000 additions a second. Computers today use transistors or microcircuits instead of tubes and can make as many as several billion computations a second. The intricacies of computers magnify their potential for misuse. Computer operation may be broken down into five principal segments—*input, programing, central process, output* and *communications.* Each is vulnerable to certain kinds of crime.

Input is the feeding of information into the computer, usually at a terminal keyboard. The input phase is vulnerable to criminal activity in two different ways. A computer operator can either introduce false data into the computer or alter the computer's records by removing vital data. The heart of the computer fraud involved in the Equity Funding case was based on

---

[6] General Accounting Office, "Computer Related Crimes in Federal Programs," April 27, 1976.

false input—that is, 56,000 phony life insurance policies. Payroll accounts are especially vulnerable to manipulation at this stage.

Programing consists of detailed instructions given to the computer to solve problems. Programs are placed in storage where the information to be processed and the rules to be used in processing it are kept until needed. Law-enforcement officials have reported dozens of cases involving illegal tampering with computer programs. Many crimes of this type occur in banks. Computer operators, bank officers or other employees with access to a bank's computer have wrongfully programed instructions to take money from accounts and transfer it to others.

The "round down fraud" is one type of programing crime. It has beeɴ used with computer systems in institutions such as large savings banks that have large numbers of financial accounts. A crime of this sort involves a computer program that rounds down an amount such as $59.11544 to $59.11, thus leaving a remainder of .00544 cents. While that figure is infinitesimal, the constant rounding down of thousands of figures adds up quickly. In normal circumstances round-down remainders are distributed to all the accounts in a bank, but a computer program can be altered to place them in a separate account —where the money can be withdrawn.

"Experienced accountants and auditors indicate the round down fraud technique has been known for many years, even before the use of computers," Donn Parker has written. "But to what extent is this done in the complex environment of computer technology?"[7] The answer is that no one knows because of the ease with which such crimes can be accomplished.

Access to Stored and Transmitted Data

Central Processing Unit is the computer's "brain" or "memory bank" where the processing of information actually takes place. It acts on instructions from the program. Writing in *Barrister,* an American Bar Association magazine, August Bequai maintains that the central processing unit is "extremely vulnerable" to "attack from wiretapping, electro-magnetic pickups or browsing."[8]

Knowledgeable persons have been able to gain access to central processing units to steal secrets, such as corporate data, or personal information. Information stolen from computers has been sold to rival companies, used for ransom or blackmail, and for setting up complicated programs without the expense of planning and writing them. Jerry Schneider, who stole nearly $1-million of telephone equipment in Los Angeles, did so by

---

[7] Parker, *op. cit.,* p. 117.

[8] August Bequai, "The Electronic Criminal," *Barrister,* winter 1977, p. 11.

tapping into Pacific Telephone & Telegraph's computer. Many crimes of this sort involve a persons working for the victimized company. But the computer tampering in Schneider's crime was carried out by Schneider alone after he surreptitiously learned several of the telephone company's computer codes. "It was very easy to do," Schneider said. "...I did it myself from the outside. I just called up on the telephone and placed the orders inside the computer."[9]

Output is the processed information provided by the computer. Output may consist of mailing lists, payroll checks or private, secret or other sensitive information. Output data, like any other valuable commodity, is subject to theft.

Communication is transference of output data between computers. It usually is accomplished by telephone or teleprinter. This aspect of computer operations is subject to electronic interception either to change or to steal data. Communications crimes are particularly difficult to detect and are attractive to criminals because the perpetrator often is far from the scene of the crime.

One such case involved an illegal transfer of funds between two banks some 3,000 miles apart. A man opened a large account at a New York bank, informing bank officials that he was a West Coast manufacturer about to open a new factory in the East. He instructed the New York bank to expect the transfer of a large sum of money from his West Coast bank to finance the new business venture. Subsequently $2-million was transferred to the New York bank by computer instruction—a standard procedure among major financial institutions.

In this instance, the "manufacturer" received his money and fled. Police investigations revealed that he had conned a woman computer operator at the West Coast bank into sending a message to transfer the $2-million in the belief she was helping him play a joke on a friend. "Since the missing...pseudo-manufacturer was not available for questioning by the police," Thomas Whiteside wrote, "the two banks involved in the two-million-dollar transfer were left with the computerized message as a souvenir."[10]

Virtual Absence of Computer Security

One reason why computer crimes occur is that there is virtually no security at most computer operations. Moreover, a foolproof method of protecting computers has yet to be developed. "Fundamentally, we do not know how to protect large-scale, multi-access computer systems," Donn Parker said

---

[9] Quoted by McKnight, *op. cit.*, p. 38.
[10] Whiteside, *op. cit.*, Aug. 22, 1977, pp. 49-50.

recently.[11] The 1976 GAO report said that the problem is the same in the federal government. Every incident of computer misuse reviewed by the agency was "directly traceable to weaknesses in system controls...the result of deficient systems designs, improper implementation of controls by operating personnel or a combination of both."

Of the computer criminals who are caught, many are tripped up only because of some unrelated activity. The head teller of the Park Avenue branch of the Union Dime Savings Bank in New York City was convicted in 1974 of embezzling $1.5-million by raiding depositors' accounts and covering up the crime by manipulating the bank's computer. The losses were not uncovered until police raided an illegal gambling emporium. They found receipts indicating that the teller bet as much as $30,000 a day on horse races and other sporting events. The teller's annual salary was $11,000. When police questioned the teller, he confessed that he had been stealing from the bank for three years.

Jerry Schneider was brought to justice only after one of his employees became angry at being refused a raise. The disgruntled employee called the police. Informants' tips have uncovered many other computer-related crimes, including the Equity Funding scandal. First word of the scandal came when a former Equity employee informed the New York State Insurance Department that he suspected wrongdoing by his former employer.

Efforts to Improve Safeguards, Training

This is not to say, though, that computer supervisors are not aware of security problems and are doing nothing about it. The Federal Bureau of Investigation recently opened a special course at its Quantico, Va., training center to train FBI agents and state and local police for investigating consumer fraud and other white-collar crimes. But the course has come in for criticism. Thomas Whiteside wrote: "...[A]ccording to commercial computer-security people I have talked with, the training so far offered is not of a very advanced type in relation to all the ramifications of the existing criminal problems."[12] August Bequai said of the FBI training program: "It's not enough to train a person on what a computer can do. You have to train him in the sophisticated and complicated frauds that a complex computer is capable of performing."

International Business Machines, the leading manufacturer of computers, recently came up with recommendations to prevent computer crime. IBM studied computer security for several years and is reported to have spent $40-million on the effort.

---

[11] Quoted in *Business Week,* Aug. 1, 1977, p. 44.

[12] Whiteside, *op. cit.,* Aug. 29, 1977, p. 40.

Volume of Annual Transactions
Affecting Data Processing

Type of Transaction	1955	1970	Increase 1955-70
Checks written	2.1 billion	7.2 billion	243%
Telephones in use	56.2 million	120.2 million	114
Individual social security payments	8 million	26.2 million	228
Individual federal tax returns	58.3 million	77 million	32
Public welfare recipients	5.6 million	13.3 million	137
Airline passengers	42 million	171 million	307
Persons entering hospitals for treatment	21.1 million	31.7 million	50
Persons covered by private hospitalization insurance	107.7 million	181.5 million	69
Motor-vehicle registrations	62.7 million	108.4 million	73
Passports issued	528,000	2.2 million	317
Students enrolled in colleges and universities	2.6 million	6.9 million*	165
Applications received for federal employment	1.7 million	2.9 million	71
New York Stock Exchange transactions	820.5 million	3.2 billion	290
Pieces of mail handled, U.S. Post Office (all classes)	55.3 billion	84.9 billion	52

SOURCE: Donn B. Parker, *Crime by Computer* (1976), pp. 242-244.
*1969

The company recommends four basic measures to help prevent computer abuse: (1) rigid physical security, (2) new identification procedures for keyboard operators, (3) new internal auditing procedures to keep a fuller record of each computer transaction, and (4) new cryptographic symbols to scramble information. Even with those and other precautions, security problems remain. "The data security job will never be done—after all, there will never be a bank that absolutely can't be robbed," IBM's director of data security, John Rankine, has said.[13]

Brandt Allen, a professor of business administration at the University of Virginia, analyzed 150 cases of computer fraud

---

[13] Quoted by Whiteside, *op. cit.*, Aug. 29, 1977, p. 58.

and concluded that most computer crime "can often be prevented by a tight system of internal control."[14] Allen recommended: (1) stricter controls of input transactions, (2) rigorous audits, (3) improved management supervision, and (4) tighter program and file controls. Controls such as those recommended by Allen and IBM have one consequence that is not popular with many business executives. They involve significant outlays of money. "Operating efficiency and [internal security] controls are counterproductive," Carl Pabst, a computer expert with the New York accounting firm of Touche, Ross & Co., said recently.[15] Pabst said that many companies do not install elaborate security systems in the hope that losses from theft will not be higher than the costs of setting up security measures.

Even the most complex security system can be broken—simply through the dishonesty of someone involved in the security itself. Donn Parker wrote, "Why go to all the trouble of technically compromising a computer center when all [one has] to do is con one of the trusted people into doing anything [one wants] him to do?"[16] One other factor compounds the problem. Most businesses and the federal government do not screen employees who work directly with computers. Nor are computer operators, programers and the like licensed or compelled to abide by a standard code of conduct. Until such a system is implemented, some experts believe, the inefficient security at today's computer centers will add to the continuing increase in computer crime.

Criminal Justice and Technology

MOST OBSERVERS of the computer crime scene agree on one thing: computer crime is growing and will continue to grow before there is any improvement. "There are going to be a lot of shocks and horror stories out there," Carl Pabst of Touche, Ross & Co. said recently, predicting more and larger computer crimes in the future.[17] One factor that has been of prime importance in the rising rate of computer crime is the inadequacy of the criminal justice system in dealing with it.

"Criminologists, lawyers, judges—the entire legal

---

[14] Brandt Allen, "The Biggest Computer Frauds: Lessons for CPA," *The Journal of Accountancy,* May 1977, p. 52.

[15] Quoted in *The Wall Street Journal,* Oct. 5, 1977.

[16] Parker, *op. cit.,* p. 282.

[17] Quoted in *The Wall Street Journal,* Oct. 5, 1977.

system—concentrate on violent crimes such as arson, rape and murder," August Bequai said recently. "But the technological revolution has hit the area of crime and the traditional legal system has proven incapable of handling it."[18] The rules of evidence make it especially difficult to try most computer crime cases, for stolen computer material often consists not of cash or securities but of a series of invisible electronic impulses.

This situation is made worse by what Bequai called "serious procedural and evidentiary handicaps built into the legal system." He maintains that the overall inefficiency of the legal system also adds to the problem. "Right now," he said, "it takes one year to dispose of a simple shoplifting case. A complicated, sophisticated computer crime can tie up the legal system for years. The situation challenges the whole legal apparatus."

Adding to the problem of prosecution is the fact that many police investigators, prosecuting attorneys, judges and juries have only scant knowledge of the simplest computer operations. This makes the uncovering of a crime, its presentation before a jury and the sentencing of a convicted criminal extremely complicated and time-consuming. Even if a computer thief gets caught, is tried and then convicted, chances are that the sentence will be a light one. One reason is that probation officers and others who prepare pre-sentencing reports on computer criminals tend to stress the non-violent nature of the crime.

Leniency for White-Collar Offenders

Most computer criminals are white and middle class and have no prior criminal record. As a result, criminals convicted of computer crimes often wind up with light sentences. Jerry Schneider, who stole some $1-million worth of telephone company equipment, was sentenced to 60 days in jail. He served 40 days and was released. The head teller of the Union Dime Savings Bank branch in New York City, who embezzled $1.5-million, was sentenced to two years in prison and served 20 months. The pre-sentencing report on one convicted computer criminal said: "Computer-program theft is the crime of the future. As such it has not been referred to this office at the felony level in prior instances. Thus, there is no ready frame of reference for determining a proper sentence."[19]

Some computer criminals have received extremely lenient treatment from their victims. "It's like rape," commented Gary Keefe, a computer expert with the accounting firm of Peat, Marwick, Mitchell & Co. "Management feels it's been beaten and doesn't want people to know about it."[20] Banks, for ex-

---

[18] Interview, Dec. 2, 1977.
[19] Quoted by Whiteside, *op. cit.*, Aug. 29, 1977, p. 42.
[20] Quoted in *The Wall Street Journal,* Oct. 5, 1977.

Cashless and Checkless in America

The cashless, checkless society may be upon us in the not too distant future. If current progress continues, most Americans will be using electronic funds transfer (EFT) systems to pay all their bills and make store purchases and bank transactions.

Today, EFT systems are in use in bank teller machines, automatic bill payment plans and direct sales in stores that require no cash, checks or credit cards. In these point-of-sale transfers, customers make purchases using nothing more than their debit cards. These cards resemble credit cards but really are automatic checkbooks. Through computers, the amount of the sale instantly is deducted from the buyer's bank account and added to the retailer's account. No cash is exchanged, no check written, no bill sent, and often no receipt given.

The largest and most common EFT systems in use today are bank-to-bank account transfers. These systems handle billions of dollars daily and do not involve the physical movement of money. Other EFT systems are used to deposit funds or withdraw cash from automated teller machines that remain in operation when the banks are closed. Bank customers usually use a combination of their checking account card and a confidential code to dial transactions. Other EFT conveniences include automatic deposits of paychecks into checking or savings accounts and automatic payment of regular bills such as rent, car payments, mortgages and insurance. Many banks will handle bill payments after receiving telephone calls from their customers.

It is predicted that eventually a national EFT system will have some 40 million computer terminals attached to thousands of computers connecting consumers and retail stores with banking functions. Computers, of course, are at the heart of EFT systems. Some of the factors that make computers vulnerable to fraud, embezzlement and theft exist in EFT systems.

But the EFT transactions are traceable, recordable and revocable. Still, adequate security measures have yet to be developed and the criminal justice system is poorly equipped to handle litigation involving EFT systems. If these cashless transactions spell the end of street robberies and traditional bank holdups, will these crimes be replaced by electronic ones?

ample, fear publicity about large thefts involving computers. The worry is that depositors will lose confidence when they learn the vulnerability of their banks to computer crime. Publicly held corporations, too, tend to keep computer-crime losses quiet. They do not want to lose the confidence of stockholders. Gerald McKnight has written: "Because of what competitors, shareholders and loan providers such as banks might do if they found out that the company's costly computer had been broached, those in the know hush up scandal after scandal."[21]

Some offenders have been merely reprimanded. Some have been fired, but without being forced to make restitution. Others have even been given good references for future employment. An extreme example came in the case of a young executive in England who, when confronted with evidence, admitted he had been stealing from his company's computer. For fear of bad publicity, the company gave the man a letter of recommendation to help him find a new job. He soon went to work for another company as executive director and proceeded to raid the new company's computer, embezzling some $2,000 a week for three-and-a-half years. For a second time, the embezzler was uncovered but not prosecuted. Again the victimized company did not ask for restitution and provided the thief with a good employment reference.

"A million dollars from a computer crime is considered respectable but not an extraordinary score."

The nature of computer crime lends itself easily to thefts of large amounts of money. Brandt Allen, in his analysis of 150 cases of computer fraud *(see p. 73)*, found that the average loss was $621,000 among corporations, $193,000 among bank and savings institutions, $329,000 among state and local governments, and $45,000 among federal government agencies. "A million dollars from a computer crime," Thomas Whiteside wrote, "is considered a respectable but not an extraordinary score."[22]

With such large sums of money involved, law-enforcement authorities have become concerned that organized crime groups may be getting involved in computer crimes. Terry Knoepp, U.S. Attorney in San Diego, Calif., said in 1976: "My guess is that any time you have large amounts of money involved...that can be transferred without any sort of audit or tracing, you're

---

[21] McKnight, *op. cit.*, p. 47.
[22] Whiteside, *op. cit.*, Aug. 22, 1977, p. 38.

Case of the Missing Boxcars

The time: October 1971. The place: Pennsylvania. The case: the disappearance of 217 boxcars owned by the bankrupt Penn Central Railroad.

FBI agents eventually found the cars, which were worth several million dollars. They were sitting on sidings owned by the La Salle and Bureau County Railroad, 100 miles west of Chicago. The L&BC is a tiny rail company that has only 15 miles of track. The boxcars were painted over to appear to be owned by the L&BC, but the Penn Central logo was visible under some of the paint.

A federal crime task force set up to investigate the incident concluded that the boxcars were sent to Illinois through manipulation of the Penn Central's computer. The task force never fully ascertained who misrouted the cars or why. But investigators strongly suspected the cars were sent on their journey by an organized crime syndicate which planned either to sell or rent the boxcars to another railroad.

going to get organized crime interested.... I think that that's certainly a potential...."[23]

Predictions of Rise in Computer Crimes

Computer crime, Donn Parker has written, "is expected to continue to rise merely from the proliferation of computers."[24] Not only will computer crime continue to rise, but new and different forms of computer abuse likely will be developed. One new field, international computer data communication, involves large corporations and enormous amounts of money. Parker wrote that unless strict measures are developed to safeguard those international transactions, "Equity-Fund" type scandals could easily occur on an international scale.

The potential for computer abuse multiplies as computers control more and more functions. Certain aspects of city subway systems, air traffic control, hospital intensive-care monitoring, and even police and fire department functions are being computerized across the nation. In addition, the 500,000 computers that Parker predicts will be in operation by 1980 will be installed in places where white-collar crimes have been prevalent in the past.

Those who study computer crime do so from a judicial, technological or security perspective. But all agree that computer crime will continue to rise in the near future. Donn Parker does not see any improvement in the situation until the early 1980s. And his prediction of improvement then is based on the

---

[23] Interviewed on "60 Minutes" (CBS-TV), Oct. 10, 1976.
[24] Parker, *op. cit.*, p. 293.

Computer Misuse in Industry and Government Among 372 Cases Studied*

Breakdown	Number of Cases	Per Cent of Totals
Banking	70	19
Education	66	18
Government	61	16
Local	22	
State	14	
Federal	13	
Foreign	12	
Manufacturing	46	12
Insurance	28	8
Computer Services	24	6
Transportation	9	2
Retail Stores	8	2
Dating Bureaus	6	2
Trade Schools	5	1
Utilities	5	1
Communications	5	1
Credit Reporting	5	1
Securities	4	1
Petroleum	4	1
Other	26	8

*Through October 1975, by SRI International

Percentages do not add to 100 because of rounding

assumption that the development of security systems will keep pace with other technological advances in computerization. Others, including Thomas Whiteside, have painted an even gloomier picture. Whiteside wrote that the designers of future computer systems "seem no more able to promise absolute solutions to problems of data security than chess players are able to foresee games in which White can never be beaten." Until a foolproof system is developed, computer crime is likely to perplex the people that computers are intended to serve.

Selected Bibliography

Books

Adams, J. Mack and Douglas H. Haden, *Social Effects of Computer Use and Misuse,* Wiley, 1976.

Ellison, J.R. and F.E. Taylor, *Where Next for Computer Security?* International Publications, 1974.

Kemeny, John G., *Man and the Computer,* Scribner's, 1972.

McKnight, Gerald, *Computer Crime,* Walker and Co., 1973.

Parker, Donn B., *Crime by Computer,* Scribner's, 1976.

Sobel, Ronald L. and Robert E. Dallos, *The Impossible Dream: The Equity Funding Scandal,* Putnam, 1975.

Van Tassell, Dennis, *Computer Security and Management,* Prentice Hall, 1972.

Articles

Allen, Brandt, "The Biggest Computer Frauds: Lessons for CPAs," *The Journal of Accountancy,* May 1977.

Bequai, August, "The Electronic Criminal," *Barrister,* winter 1977.

—"White Collar Crime: The Losing War," *Case & Comment,* September-October 1977.

Cameron, Margaret, "Crime In, Crime Out," *Canadian Banker,* May-June 1977.

Computerworld, selected issues.

Congressional Quarterly Weekly Report, Sept. 17, 1977, pp. 1955-56.

Davis, James R., "Computer Controls—A Different Emphasis," *The National Public Accountant,* June 1977.

Infosystems, selected issues.

"The Growing Threat to Computer Security," *Business Week,* Aug. 1, 1977.

Turn, Rein, "Privacy Protection and Security in Business Computer Systems," *Atlanta Economic Review,* November-December 1976.

Whiteside, Thomas, "Dead Souls in the Computer," *The New Yorker* (2 parts), Aug. 22, 29, 1977.

Reports and Studies

Chamber of Commerce of the United States, "Handbook on White Collar Crime," 1974.

Editorial Research Reports, "Approach to Thinking Machines," 1962 Vol. II, p. 537; "Reappraisal of Computers," 1971 Vol. I, p. 347; "Crime Reduction: Reality or Illusion," 1977 Vol. II, p. 537.

General Accounting Office, "Ways to Improve Management of Federally Funded Computerized Models," Aug. 23, 1976.

—"Computer-Related Crimes in Federal Programs," April 27, 1976.

U.S. Senate Governmental Affairs Committee, "Computer Security in Federal Programs," Feb. 2, 1977.

—"Problems Associated with Computer Technology in Federal Programs and Private Industry," June 18, 1976.

Strategies for Controlling Cancer

by

William V. Thomas

Aug. 5
1977

Editor's Note: The National Cancer Institute, in September 1978, announced plans to spend $250,000 testing the controversial anti-cancer drug Laetrile on humans. The six-month trial, scheduled to begin in January 1979, will involve at least 300 patients with advanced cancer. NCI director Arthur Upton said he hoped the trial would "settle the issue" of Laetrile's effectiveness.

STRATEGIES FOR CONTROLLING CANCER

SINCE AUG. 5, 1937, when Congress established the National Cancer Institute, the federal government has made the conquest of cancer an overriding national health goal. From its inception, the cancer program's ambitious objective was to eliminate all forms of the dreaded disease. But to many people the progress of research has seemed agonizingly slow; popular enthusiasm for the unproven anti-cancer drug Laetrile *(see p. 95)* is one indication of the widespread disenchantment with the effectiveness of traditional medical knowledge.

In recent years, under mounting public pressure to match the dramatic achievements in other areas of science and medicine, legislators have poured billions of dollars into the effort to defeat cancer. Yet despite the great investment of federal money and private donations, a cure for the most frightful of all human afflictions has remained stubbornly elusive.

Today, the problems in eradicating cancer are not all confined to the medical laboratory; controversy surrounds the question of who should wage the so-called war on the disease and exactly how it should be carried out. During three days of congressional hearings in June before the House Subcommittee on Intergovernmental Relations and Human Resources, a recurring theme in the testimony was that present policies have been "largely unsuccessful"[1] and that more should be done to identify cancer's possible causes instead of trying to find its remedy.

In the past decade, there have been remarkable strides in the treatment of certain types of cancer; leukemia and Hodgkin's disease *(see glossary, p. 85)* have both responded favorably to new drug therapies. These advances, however, have tended to be offset by the rising incidence of respiratory malignancies and other forms of the disease, which many health experts attribute to environmental sources. Although most scientists concur in the belief that measures to curb human exposure to suspected toxins would produce lower mortality rates, few can agree on what portion of the cancer effort should be devoted to research in preventive medicine. The result has been a philosophical and political split in the once-unified national program.

---

[1] So stated by Sidney M. Wolfe of the Public Citizen's Health Research Group before the subcommittee, June 14, 1977.

The quest to unravel the mysteries of cancer is complicated by the very nature of the disorder. There are, in fact, more than 100 varieties of the disease, and they attack dozens of sites in the body. Once relatively uncommon, cancer is now this country's second leading killer, taking nearly 366,000 lives in 1975, the latest year for which official figures are available. That total represents about half the number of people who died of heart attacks in the same year, but twice as many as died of stroke. According to Dorothy P. Rice, director of the National Center for Health Statistics, who cited these findings in recent congressional testimony, of the three major causes of death in America, "cancer is the only one for which we have yet to see a steady, long-term decline."[2] She reported that the age-adjusted[3] cancer death rate rose from 116.7 per 100,000 population in 1930 to 130.9 in 1975. Provisional figures for 1976 show a further rising trend.

Opinion is divided on the issue of how to combat the cancer epidemic. Dr. Michael B. Shimkin, former president of the American Association for Cancer Research, has argued that the government should increase its support of the effort to discover a cure. "No responsible scientist," Shimkin writes, "has ever...thought of cancer as an insuperable problem, beyond the reach of scientific research... The national investment in solving the cancer problem is thoroughly sound, and if it proceeds without interruption, it will accomplish its aim."[4] But a growing number of persons, among them some members of Congress, disagree. They argue that the war on cancer is being fought with "outdated strategy"[5] which may never produce victory.

Controversy Over Allotment of Resources

When the National Cancer Act took effect five years ago, President Nixon declared that money would be no object to final victory over the disease, signaling the beginning of what he called a "war on cancer." "Cancer is a scourge we must fight," Nixon said. "That fight deserves, from all of us, all the money, all the resources, and all the ingenuity that are required to win it."[6] President Carter similarly made commitments of dollars and faith to solve "our greatest medical problem."[7] Of the more than $2-billion in federal funds made available to the cancer program during the last 40 years, 62.6 per cent has been spent

---

[2] Testimony before the House Subcommittee on Intergovernmental Relations and Human Resources, June 14, 1977.

[3] Age-adjustment compensates for the increased number of deaths that occur in old age.

[4] Michael B. Shimkin, "The Coming Cure for Cancer," *Skeptic*, May-June 1977, p. 58.

[5] Phrase used by Daniel S. Greenberg and Judith E. Randal, "Waging the Wrong War on Cancer," *The Washington Post*, May 1, 1977. Greenberg is editor of *Science and Government Report* newsletter. Randal reports on science and medicine for the New York *Daily News*.

[6] From a speech by Nixon to the National Cancer Conference in Los Angeles, Sept. 28, 1972.

[7] So characterized by Carter, March 8, 1977, in declaring March "Cancer Control Month."

Glossary of Cancer Terms

Cancer. A general term, derived from the Latin word meaning crab, used to indicate any of various types of malignant diseases that invade the body and are likely to result in illness or death.

Carcinogen. A term used to describe any cancer-producing agent.

Carcinoma. A solid malignant tumor that originates in the skin, glands, nerves, breasts, lungs, stomach, digestive and urinary tracts. Carcinoma accounts for 80 to 90 per cent of all cancer cases in the United States.

Leukemia. Cancer of the blood-producing organs caused by the overproduction of immature white corpuscles; it accounts for about 4 per cent of all U.S. cancer cases.

Lymphoma. A form of cancer resulting from an abnormal production of immature lymphocytes by the spleen and lymph nodes; **Hodgkin's disease,** the most common type of lymphoma, accounts for about 3 per cent of all cancers.

Sarcoma. A solid malignant tumor growing in muscle, bone, cartilage and connective tissue; it accounts for only about 2 per cent of the U.S. cancer cases.

Tumor. A mass of abnormal tissue arising from pre-existing cells, serving no known purpose and growing independently of surrounding tissue. Benign tumors cause damage only when they interfere with the function of an organ. Malignant tumors kill by causing generalized emaciation and ill health in the host. They do not produce toxins, but take a priority on the body's nutrients, ultimately causing infections and organ failures.

since 1972.[8] The Carter administration budget for 1978 requests $819-million to fight the disease.

The role of government as the preeminent sponsor of biomedical research is firmly established. But at a time when public agencies are being called upon to account for how they spend federal money, the fact that the cancer enigma remains largely unsolved has moved some critics to suggest that there is fiscal mismanagement in the national cancer program and, to quote one, it is "warping scientific motivation."[9] Guy R. Newell, acting director of the National Cancer Institute, defends the program. He told Congress recently: "The American people must temper their expectations with the inescapable realization that cancer encompasses a host of very complex and different diseases that yield reluctantly to new knowledge."[10]

The President's Biomedical Research Panel, after completing a 15-month study in 1976, advised the White House that the National Cancer Institute and its parent agency, the National

[8] Figure cited by Guy R. Newell, acting director of the National Cancer Institute, in testimony before the House Subcommittee on Intergovernmental Relations and Human Resources, June 15, 1977.

[9] Solomon Garb, director of the American Cancer Research Center, testimony before the House Subcommittee on Intergovernmental Relations and Human Resources, June 14, 1977.

[10] Testimony of Guy R. Newell (see footnote 8).

Institutes of Health, would function best if left under the direction of the scientific community. The federal health effort as a whole, the panel concluded, should have a steady infusion of money and less interference by politicians.[11] The panel reported that since 1971 the cancer institute had spent $2.3-billion on research, representing 32 per cent of all federal spending on health research.

In general, the feeling was that the money had been spent well. However, economy-minded members of the House Appropriations Committee, currently surveying the operation of the cancer program, contend that the study of cancer needs closer oversight. Rep. David R. Obey (D Wis.), a committee member, said the time has come for NCI "to explain how it spends its millions."[12] Although other government agencies are involved in cancer research, NCI, through its support of regional treatment centers across the country and the awarding of nearly 2,500 research grants each year, controls more than 90 per cent of all federal anti-cancer funds.

There have been indications that some cancer officials might also support stricter legislative monitoring. R. Lee Clark, president of the American Cancer Society, a private fund-raising organization, told Congress that increased accountability could help to coordinate the diverse range of federally supported research projects.[13] But the problem, all agree, will be finding acceptable criteria for measuring accountability in an area of medical investigation where progress up to now has been painstakingly slow.

Questioning the Stress on Viral Research

After the conquest of polio in the 1950s, it was hoped that massive injections of federal money into the national drive to defeat cancer would hasten the arrival of the day when all forms of the deadly disease might be prevented by a single inoculation. "It is not surprising that the common view, now decades old, is that a cure for cancer must certainly be within our grasp," author Larry Agran writes in *The Cancer Connection*. "But the grim reality is otherwise. It is unlikely that there will be any sudden breakthrough in the near future. There will be no universal cure.... Probably not in this century. And possibly never."[14]

When the federally proclaimed war on cancer began in 1972, viral research was thought to hold the key to the development of an effective anti-cancer vaccine. However, in the entire history of

---

[11] "Report of the President's Biomedical Research Panel," April 30, 1976, pp. 25-32.
[12] Quoted in *Congressional Insight*, June 4, 1977, p. 4. *Congressional Insight* is a weekly newsletter published by Congressional Quarterly Inc.
[13] Testimony before the House Subcommittee on Intergovernmental Relations and Human Resources, June 15, 1977.
[14] Larry Agran, *The Cancer Connection, And What We Can Do About It* (1977), p. xxi.

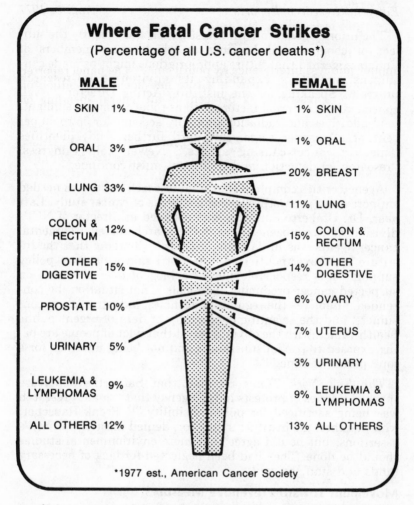

Where Fatal Cancer Strikes
(Percentage of all U.S. cancer deaths*)

MALE			FEMALE
SKIN	1%	1%	SKIN
ORAL	3%	1%	ORAL
		20%	BREAST
LUNG	33%	11%	LUNG
COLON & RECTUM	12%	15%	COLON & RECTUM
OTHER DIGESTIVE	15%	14%	OTHER DIGESTIVE
PROSTATE	10%	6%	OVARY
		7%	UTERUS
URINARY	5%	3%	URINARY
LEUKEMIA & LYMPHOMAS	9%	9%	LEUKEMIA & LYMPHOMAS
ALL OTHERS	12%	13%	ALL OTHERS

*1977 est., American Cancer Society

medicine not a single human cancer virus has been isolated. "All animals have been proven to get cancer from viruses," said R. Lee Clark. "Why man should be exempt, we don't know."[15]

In 1975, the presidentially appointed Cancer Advisory Board reported that finding a viral cause for most human cancers was "an unlikely eventuality" and that an "emphasis on viral oncology [cancer study] should perhaps be reconsidered." But the two major cancer-fighting organizations, the National Cancer Institute and the American Cancer Society, have continued to invest heavily in viral research. While interest in tracking down environmental carcinogens has grown in the last several years, NCI, the scientific command post of the cancer war, committed only about 15 per cent of its 1977 budget to study in that field.

[15] Quoted in *Washington Drug and Device Letter*, June 20, 1977, p. 1. The newsletter is published by the Health News Services Division of Capitol Publications Inc., Washington, D.C. See also "Virus Research," *E.R.R.*, 1970 Vol. I, pp. 621-640.

The priority given to viral investigation was recently the subject of congressional hearings at which some members of Congress warned that future appropriations might be made contingent on NCI's expanding its environmental-research program. Speaking for the institute, Acting Director Newell conceded that external factors such as tobacco smoke, sunlight, food additives and radiation probably account for 55 to 60 per cent of all cancer cases. But until further study identifies "specific cancer-causing agents," Newell told Congress, "research on detection and treatment must continue."

Agency critics complain that viral research has drawn needed support away from other important areas of cancer study. Last year, Dr. Umberto Saffiotti, who resigned as director of NCI's division of carcinogenesis, said the search for environmental cancer agents was in "a state of crisis." Charging that the institute was "over-politicized," Saffiotti said a "lack of policy support" was responsible for a backlog of over 200 tests on suspected cancer-producing chemicals. That situation, he contended, "deprives the regulatory agencies, industry, labor, consumers, and the scientific community of data of urgent public health value...with the obvious result that a lot of people are being exposed to [toxic] substances and not being informed for a long period of time."[16]

The New York Times reported that Saffiotti's departure sparked a series of protests in the agency that "quality research was being sacrificed for public visibility."[17] Frank Rauscher, who was then the institute's director, denied most of Saffiotti's assertions, but he did agree that more envirionmental studies should be done. They had been neglected for lack of necessary funds and staff, he said.

Movement Toward Preventive Measures

Theories guiding cancer research have undergone a number of important changes during the last four decades. In the 1940s, there was intense interest in radiation as a cancer-inducing agent. In the 1950s and 1960s, attention was focused on genetic causes. Lately, many scientists consider chemical carcinogens in the environment to be the primary source of human cancer. That notion received endorsement last year from Russell Train, who was then director of the Environmental Protection Agency. He said:

"Most Americans had no idea until relatively recently that they were living so dangerously...that when they did things as ordinary, and as essential to life as eat, drink, breathe or touch,

---

[16] Quoted in *Environmental Action*, June 4, 1977, p. 13.
[17] April 30, 1976.

they could in fact be laying their lives on the line. They had no idea that, without their knowledge or consent, they were often engaging in a grim game of chemical roulette whose results they would not know until many years later."[18]

As new ideas about cancer risks continue to gain acceptance, sentiment within the scientific community seems gradually to be shifting toward cancer prevention rather than treatment. Advocates of prevention concede that it would encounter difficulties—to avoid the potential causes of cancer Americans would have to abandon many of their traditional dietary, cultural and social habits, just as doctors and medical researchers would have to give up their orientation toward curative investigation.

According to the Federation of American Scientists, if preventive measures could reduce the U.S. rate for each type of cancer to the lowest rate observed in the world, this country would eventually experience a 90 per cent decrease in its cancer deaths.[19] An essential part of the prevention effort is the testing of thousands of new chemicals introduced each year. Short of exposing human subjects to lethal substances, animal experimentation is the usual procedure for determining toxicity. Potent carcinogens tend to become evident early in testing. But gauging the long-term dangers of a suspicious substance can be slow and painstaking. The results are often a matter of dispute, since duplicating the exact conditions of human exposure are considered to be nearly impossible. Moreover, most forms of cancer are believed to have an average incubation period of 20 to 30 years, which makes precise connections between cause and effect hard to establish.

Researchers are hopeful that laboratory experiments like the Ames test,[20] which measures the cancer-producing effect of chemicals on a bacteria culture, will eventually replace time-consuming animal studies. Laboratory screening procedures may also prove to be useful substitutes for the X-ray method of cancer detection, which many doctors believe needlessly subjects persons being tested to potentially harmful doses of radiation.

One of the thorniest issues in cancer testing involves the question of reliability. Can any lab test accurately predict the effect of a suspected carcinogen on humans? Some toxicologists contend that the use of animal data may unfairly implicate certain

---

[18] Quoted in *The Washington Star,* May 23, 1976.

[19] Figures cited in *F.A.S. Public Interest Report,* May 1976, a newsletter published by the Federation of American Scientists. The federation describes itself as a lobbying group of 7,-000 natural and social scientists and engineers "who are concerned with problems of science and society."

[20] The test was developed by Dr. Bruce N. Ames, professor of biochemistry at the University of California at Berkeley.

chemicals. The only reliable information on cancer hazards for humans, they argue, can be drawn from direct human experience. Dr. Richard Griesemer, director of the National Cancer Institute's carcinogenesis bioassay program, disagrees. "When a chemical is a carcinogen in man, it's also a carcinogen in animals," he said. "So we have to assume that every chemical hazardous in animals is also a carcinogen in man until proved otherwise."[21]

Cancer and the Environment

R EPORTS of cancer fatalities from environmental agents have become as familiar today as Vietnam death statistics were a decade ago. Gone are the names of foreign battlefields, replaced by such odd-sounding chemical compounds as tris, vinyl chloride, DES. There is uncertainty about the proportion of cancer cases brought on by carcinogens—federal health officials suggest it is between 55 and 60 per cent, while other estimates run as high as 85 per cent. But there seems to be little doubt in the minds of most experts that exposure to toxins takes a sizable toll.

Studies of immigrant populations lend support to the belief in environmental causes of cancer. People who migrate to a country as children tend to take on the cancer patterns of that country, whereas people who migrate as adults usually retain the patterns of their native land. In Japan, for example, there is a very high incidence of stomach cancer. But Japanese adults who came to America as children experience the disease at a rate 40 per cent lower than people living in Japan—even though the susceptibility of Japanese-Americans is still high by U.S. standards.[22]

According to estimates by the National Cancer Institute, practices that can be controlled by humans, such as smoking and alcoholic drinking, account for one-third of this country's cancer deaths. Curbing alcohol intake and tobacco use alone could result in 75,000 lives saved each year, said Martin Schneiderman, director of field studies and statistics for NCI. However, it is easy to list the number of deaths that could be prevented, he added, and "almost impossible to get the American people to change their bad habits."[23]

---

[21] Quoted in *The Washington Post,* July 19, 1977.
[22] National Institutes of Health, *Atlas of Cancer Mortality Among U.S. Nonwhites: 1950-1969,* January 1977, p. viii.
[23] Quoted in *The Washington Post,* March 30, 1976.

Concentrations of Throat and Lung Cancer

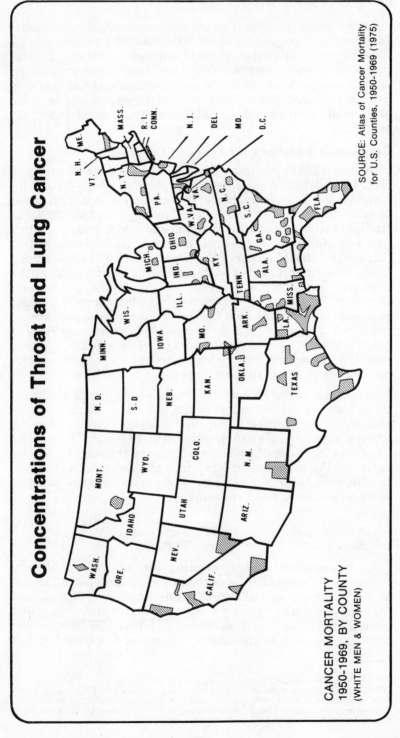

SOURCE: Atlas of Cancer Mortality
for U.S. Counties, 1950-1969 (1975)

CANCER MORTALITY
1950-1969, BY COUNTY
(WHITE MEN & WOMEN)

Today, there are more than one million chemical compounds. That total is increased by about 700 annually. Of the nearly 1,-600 toxic chemicals listed by the Department of Health, Education and Welfare, around 600 have been demonstrated to produce cancer in animals.[24] Many scientists theorize that it is these substances, some of which most Americans use or come into contact with every day, that present the greatest dangers to health.

Geographic Variations for Types of Cancer

In June 1975 and in January 1977, the National Institutes of Health published nationwide studies showing county-by-county variation for 35 types of cancer.[25] Based on information taken from death certificates and compiled for 20 years, 1950-1969, the studies provide clues to occupational and environmental factors that may contribute to the onset of cancer. NIH scientists identified high rates of cancer of the lung, liver and bladder among males in geographic areas with significant employment in the chemical industry, specifically citing Buffalo, Chicago, Cleveland, Detroit and Milwaukee. High incidence of bladder and lung cancer for males in the Northeast may also indicate occupational influences. Statistics reveal above-average rates of lung cancer for both men and women, white and black, where smelting operations are located. Arsenic, a known cancer-causing agent, is an airborne byproduct of smelting.

In all areas of the country, blacks showed higher rates of mortality than whites for cancer of the mouth, stomach, prostate and cervix. In addition, other NIH findings revealed that the survival rate for black cancer patients is 40 per cent lower than for whites. Statistically, white women have the best chances for surviving the disease, and black men the worst, as is shown in the following table of persons who were reported to be living 10 years after cancer was detected:[26]

| White women | 42% | White men | 25% |
| Black women | 33% | Black men | 15% |

An independent study indicating the influence of dietary habits on life expectancy was conducted by Dr. James Entrom of the University of California at Los Angeles. He found that Mormons have an extremely low rate of cancer mortality. Their church—the Church of Jesus Christ of Latter-Day Saints—forbids its members to use coffee, alcohol and tobacco, and it

[24] Figures cited in *Environmental Action*, June 4, 1977, p. 13.
[25] National Institutes of Health, *Atlas of Cancer Mortality for U.S. Counties: 1950-1969*, June 1975, and *Atlas of Cancer Mortality Among U.S. Nonwhites: 1950-1969*, January 1977.
[26] National Institutes of Health, *Treatment and Survival Patterns for Black and White Cancer Patients: 1955-1964*, 1975, p. 3.

Milestones in Cancer Research

1775. The first description of occupation-caused cancer was recorded by Percivall Pott of England, who related the soot in smokestacks with cancer of the skin and scrotum among chimney sweeps.

1840. Microscopic studies by Johannes Muller of Germany first established the cellular nature of cancer.

1877. Cancer tissue successfully transplanted in animals by Russian scientist M. Novinsky provided the first experimental material for research.

1898. The first American laboratory designed exclusively for the study of cancer was opened in Buffalo, N.Y., under a $10,000 appropriation by the state legislature.

1910. The viral theory of cancer was first advanced by Peyton Rous of the United States, who discovered that the disease could be transmitted among fowls by the grafting of cancerous tissues.

1915. The search for a carcinogenic agent in tar was begun by K. Yamagiwa of Japan, who induced cancer in rabbits by the application of coal tar to their ears.

1942. Chemotherapy was first used by American physician Charles Huggins in the hormone treatment of prostate cancer.

Early 1960s. Techniques of modern immunology and molecular biology were first applied to the treatment of cancer by Swedish and American investigators.

cautions against the excessive consumption of meat. Utah, where Mormons make up nearly three-fourths of the population, has a cancer death rate 25 per cent below the national average.

Federal Commitment to Reduce Cancer Risks

It is the task of a host of federal regulatory agencies to oversee a bewildering variety of laws that govern human exposure to toxic substances. The Environmental Protection Agency (EPA) tests thousands of potentially harmful pesticides and monitors the emission of pollutants into the air and water. The Occupational Safety and Health Administration (OSHA) sets and enforces standards to protect workers from on-the-job hazards. The Food and Drug Administration (FDA) issues health regulations on more than 2,500 food additives, 4,000 cosmetics and hundreds of thousands of commercial drugs.

The government's role in protecting people from dangerous substances began at the turn-of-the-century when enthusiasm for public health legislation swept the country. The FDA's mandate goes back to 1906. But it has only been in the last 20 years[27]

---

[27] One of the most important revisions of regulatory powers came in 1958 when Congress approved the Delaney Amendment of the Food Additive Act. The measure, named for its chief sponsor, Rep. James J. Delaney (D N.Y.), gave the FDA the authority to remove from the market any food additive found to cause cancer in animals.

that concern over the pervasive presence of carcinogens in the environment prompted the expansion of regulatory functions to include cancer investigation. Lately, that aspect of the agencies' activity has been the focus of considerable controversy. Opponents of regulation argue the agencies tend to overreact in favor of consumer interests, while pro-regulation spokesmen often accuse them of inaction and a reluctance to use their full power. The agencies, caught in the middle, contend they are working to their capacity under extremely difficult circumstances.

Since its creation in 1970, the EPA has banned three pesticides from sale in the United States: DDT *(dichloro-diphenyl-trichloroethane)* in 1972, aldrin in 1974 and most recently chlordane. The Occupational Safety and Health Administration, which came into existence in 1971, has set standards for asbestos, vinyl chloride and 14 other suspected carcinogenic agents. Between 1950 and 1974, the FDA removed at least 17 cancer-producing substances[28] from the food supply. In what may prove to be its most controversial action, the agency, in March, announced plans to ban a widely used sugar substitute, saccharin, after Canadian tests revealed it caused bladder cancer in laboratory rats. The proposed ban provoked a storm of opposition among dieters, diabetics and beverage manufacturers. The ban may be rescinded by Congress before it becomes effective.[29]

The removal of a suspicious chemical or drug from the market typically entails a long process, sometimes involving years of hearings, lawsuits and public debate. Agency officials have said repeatedly that political considerations play no part in their decision-making. Much of the public seems to remain unconvinced. Moreover, there is a question of whether political considerations are necessarily bad—they may represent special interests but at times they may also represent the public's interests. National health matters, it is said, are too big and important to be left solely in the hands of the medical and scientific community.

Barriers to Progress: Cost and Delivery

What this nation spends on health care—$139-billion in 1976 by official reckoning, some 7 per cent of the gross national product—is exceeded only by its spending on construction and agriculture. According to a forecast by the Department of Health, Education and Welfare, health spending could amount to 10 per cent of the gross national product by 1980. "With little reason to be cost-effective," said HEW Secretary Joseph A.

---

[28] Outlawed food colors were red dye number 1, 2 and 3, green number 1 and 2, violet number 1, yellow number 1, 2, 3 and 4, orange number 1 and 2, and carbon black; outlawed additives were cyclamates, dulcin, oil of calamus and coumarin.
[29] For background, see "Obesity and Health," *E.R.R.*, 1977 Vol. I, p. 461.

Califano Jr., "it is no wonder that our present health-care system is recklessly consuming resources that could be used to meet other needs." Speaking at the American Medical Association's 1977 convention in June, Califano added:

> Today the leading killers are not communicable [illnesses] but heart disease and cancer which are often caused by people's lifestyles and the hostile influences of the environment. Yet our system is not really geared to the field of prevention aimed at these killers. What we lack is sufficient will and ingenuity in educating our fellow citizens about how to live more safely and sensibly. And what we sorely lack is a health-care system with economic incentives as strong to prevent as they now are to cure.

The cost of cancer is high, both in economic terms and in human suffering. President Carter has pledged his administration to reduce individual medical expenses and to take steps to insure that federal health funds are used in the most effective way possible. Califano said plans for cancer research and patient care include (1) stressing prevention and early treatment, and (2) taking steps to increase competition among private drug companies and laboratories.

Medical organizations have accused the government, specifically the FDA, of retarding progress against cancer by delaying the release of anti-cancer drugs. Some cancer specialists have decried the "drug lag"—the lapse of time before drugs approved for use abroad gain official acceptance in this country where stricter regulatory standards prevail. Two drugs currently at issue are dimethyl sulfoxide (DSMO) and Gerovital.

DSMO, a chemical byproduct of papermaking, is widely prescribed in Europe, Canada and Australia to heal sores and bruises and reduce the pain associated with cancer. Gerovital, a chemical compound similar to the painkiller Novocain, is marketed in Europe and used by cancer patients. The FDA has declared that Gerovital is neither effective nor safe; DSMO can be used legally in the United States only as a veterinary medicine. In June, the American Medical Association asked the federal government to coordinate its drug testing with other countries to speed access to useful cancer medication.

Focuses of Popular Discontent

MANY VICTIMS of cancer and their desperate families are insisting they have the right to use whatever cancer remedies they wish. The immediate object of their anger is the medical "establishment," which they consider engaged in a "conspiracy" with government regulatory agencies to deprive them of drugs and treatment they believe would be effective in curing cancer. This thinking has found expression in the campaign to revoke the FDA's ban on the sale and interstate shipment of Laetrile, a drug that the agency has declared worthless as a cancer cure.

Apparently heedless of official warnings that to leave orthodox treatment is to reject the only chance of surviving cancer, proponents of patients' rights have carried their fight against proscribed therapies and drug bans to Congress where a number of "freedom of choice" bills have been introduced. One measure, offered by Rep. Steven D. Symms (R Idaho), proposes that the question of disease treatment be left entirely to the determination of patients and their doctors. Rep. Larry P. McDonald (D Ga.), a physician and one of the hundred or more co-sponsors of the Symms bill, said: "Most people don't want government medicine."[30] He said thousands of Americans have left the country in search of cancer remedies they think will help them.

Before World War II, the bulk of medical research in this country was privately financed. Today, taxpayers support a large part. This change has opened medical policymaking to increased public scrutiny, and for legislators it has raised ticklish questions of when and how far to intercede. Perhaps the knottiest moral question for all is at what point do potential dangers to the public welfare preclude the individual's right to exercise self-determination in matters of health?

Medical Community vs. Drug's Backers

The controversy surrounding the effectiveness of Laetrile makes this question a matter of life or death for some. Those who favor removal of the FDA's ban against Laetrile claim the drug prevents cancer and controls its spread. Laetrile contains the cyanide extract of fruit pits which, according to the product's advocates, poisons diseased cells while sparing normal ones. Normal cells are said to possess a neutralizing enzyme. Laetrile promoters, many of whom have been arrested for smuggling the substance into the United States from Mexico where it is made legally, point to numerous patient testimonials attesting to its cancer-curing powers.

The FDA, however, has held to its opinion that Laetrile is of unproven medical worth. In a recent bulletin to physicians around the country, the agency said: "It would be contrary to the public interest to exempt Laetrile, as some propose, from the efficacy requirements of federal law (the Kefauver-Harris

---

[30] Quoted in *Congressional Quarterly Weekly Report*, July 2, 1977, p. 1347.

Krebiozen—Yesterday's Laetrile Debate

In the early 1960s, the controversial drug krebiozen received wide publicity as a cancer cure. Although the American Cancer Society and the American Medical Association both reported there was "no basis to believe" krebiozen, derived from horse blood, had any medical value, thousands of supporters, including the late Sen. Paul Douglas (D Ill.), rallied to its defense.

Records produced by Dr. Andrew C. Ivy, then vice president of the University of Illinois and krebiozen's chief spokesman in this country, purported to show that the drug had arrested cancer in 51 per cent of a group of test patients. Numerous other cancer victims claimed krebiozen had saved them from death.

The Food and Drug Administration repeatedly sought from Dr. Ivy exact data on clinical testing of krebiozen. But the information was never produced. Finally, the krebiozen debate died out in 1965 when Ivy and other promoters of the drug were prosecuted for fraud and acquitted. Later the group was found guilty of jury tampering.

Amendments).[31] Such an exemption would set an unacceptable precedent for other unproven drugs. The 'evidence' of efficacy presented by Laetrile promoters consists entirely of hearsay arguments and patients' testimonials. The FDA and the National Cancer Institute have reviewed 'success stories'...and failed to find evidence of therapeutic effect."[32] FDA officials and doctors are also concerned that a retreat on Laetrile "would serve as a dangerous precedent, opening the way for pressure group politics on behalf of other unproven remedies."[33]

In defiance of the federal ban, Laetrile supporters, under the coordination of the California-based Committee on Freedom of Choice in Cancer Therapy, have successfully persuaded 11 state legislatures to pass laws legalizing the drug.[34] More states are considering similar measures, while elsewhere courts have allowed some individuals to receive Laetrile from their doctors. At the very least, advocates contend, the substance cannot harm dying patients, and therefore should be made available to them.

During hearings in July before the Senate Subcommittee on Health and Scientific Research, headed by Sen. Edward M. Kennedy (D Mass.), several witnesses suggested Laetrile's promise was a cruel deception. "The proponents of Laetrile," said Dr. Joseph F. Ross, professor of oncology at the UCLA

---

[31] Passed in 1962, the Kefauver-Harris Amendments to the Food, Drug and Cosmetic Act of 1938 set rules governing the safety of testing drugs on humans. They also stipulate that the FDA should not interfere with the innovative use of licensed drugs by physicians.
[32] *FDA Drug Bulletin*, January-April 1977, pp. 3-4.
[33] Jonathan Spivak, writing in *The Wall Street Journal*, July 21, 1977.
[34] Arizona, Alaska, Delaware, Florida, Indiana, Louisiana, Nevada, New Hampshire, Oklahoma, Texas and Washington.

medical school, "have stated that it is harmless. This is contrary to fact." Ross gave the subcommittee evidence purporting to show that at least 17 deaths in this country last year were attributable to Laetrile and cyanogenic (fruit kernel) poisoning. Government-sponsored testing of Laetrile on animals has so far failed to bear out claims that it prevents or cures cancer. After a series of studies, the Battelle Memorial Institute in Columbus, Ohio, reported the drug had "no effect" on a strain of human cancer induced in laboratory animals.

Speculating that controversy over Laetrile will not be put to rest until human tests are performed, Senator Kennedy in late July urged government health officials to monitor the use of the drug on large numbers of cancer patients in order to learn its real worth one way or the other. But opposing parties in the Laetrile debate have been unable to agree on what kind of testing procedures would be fair.

Cautious Faith in Pioneering Treatments

For decades, the general treatment for cancer has consisted mainly of surgery, radiology and chemotherapy. Recently, however, pioneering innovations in other treatment methods along with new combinations of traditional techniques have been tested successfully and seem to be gaining increased acceptance among patients as well as some physicians.

While many victims of cancer, sometimes in desperation, have resorted to unproven remedies to combat the disease, others have availed themselves of the latest medical therapies. The results in a few of these cases have provided researchers with hopeful signs that science may be making inroads against certain types of cancer. The drug adriamycin has been shown to have a significant effect in treating soft tissue and bone cancers. Italian doctors have found that the recurrence of breast cancer can be drastically cut by three drugs: cycophosphamide, methotrexate and fluorouracil (CMF).

Pimeson therapy, using miniature atomic explosions within cancer cells, also has been proven in tests to be effective in controlling the spread of the disease. By far the most promising discoveries involve immunotherapy, a method for getting the body to cure itself of cancer. Injections of tuberculin extract, for example, have worked in certain cases to mobilize the body's own natural immunizing defenses against cancer cells. Some medical authorities have termed these new developments "nothing short of spectacular;"[35] however, most physicians and researchers, cautious about forecasting breakthroughs, are calling for intensified study in areas where advances have been made.

[35] Editorial in *The New England Journal of Medicine*, Feb. 19, 1976, pp. 440-441.

As encouraging as many of these discoveries may seem, critics of the national cancer program contend too little money is spent exploring scientific leads that run counter to prevailing research trends. They assert that the National Science Foundation and the National Institutes of Health, in relying on the scientific community for guidance, have grown too conservative. Daniel S. Greenberg writes that bureaucracies that govern research are "concerned more with security than science."[36] Long-shot experiments with limited support have rarely received establishment funding. "It might be useful," Greenberg writes, "to respect scientific eccentricity. Out of the $24-billion that we're now spending on research, there ought to be a few million to prevent science from becoming stereotyped." The history of great medical discoveries is largely the history of risk-taking. And it is this important aspect of scientific inquiry, nurtured outside the mainstream, that may eventually lead man to final victory over cancer.

[36] Daniel S. Greenberg, writing in *The Washington Post,* July 19, 1977.

Selected Bibliography
Books

Agran, Larry, *The Cancer Connection, And What We Can Do About It,* Houghton Mifflin, 1977.

Hixon, Joseph, *Patchwork Mouse,* Doubleday, 1976.

Illich, Ivan, *Medical Nemesis: The Expropriation of Health,* Pantheon, 1976.

Randall, Willard S. and Stephen D. Solomon, *Building 6: The Tragedy at Bridesburg,* Little, Brown, 1977.

Richards, Victor, *Cancer, The Wayward Cell: Its Origins, Nature and Treatment,* University of California Press, 1972.

Articles

"American Way of Cancer," *The Economist,* June 25, 1977.

Bonham, George W., "Is American Science Good Enough?" *Change,* June 1977.

"Cancer: Running the Gauntlet for Twenty-five Years," *F.A.S. Public Interest Report,* May 1976.

Johnson, G. Timothy, "What Everyone Should Know about Cancer: A Program of Personal Preventive Medicine," *Harvard Magazine,* July-August 1977.

Journal of the National Cancer Institute, selected issues.

Scott, Rachel, "Asbestos: Can We Get Away from It?" *Environmental Action,* March 26, 1977.

Sherman, John F., "The Organization and Structure of the National Institutes of Health," *The New England Journal of Medicine,* July 7, 1977.

Reports and Studies

Atlas Of Cancer Mortality For U.S. Counties: 1950-1969, National Institutes of Health, 1975.

Atlas Of Cancer Mortality Among U.S. Nonwhites: 1950-1969, National Institutes of Health, 1977.

Editorial Research Reports, "Job Health and Safety," 1976 Vol. II, p. 951; "Quest for Cancer Control," 1974 Vol. II, p. 623; "Cancer Research Progress," 1967 Vol. I, p. 221.

"Epidemiologic Transition in the U.S.," *Population Bulletin,* Population Reference Bureau Inc., May 1977.

"Report of the President's Biomedical Research Panel," U.S. Department of Health, Education and Welfare, April 30, 1977.

Genetic Research

by

Sandra Stencel

**Mar. 25
1 9 7 7**

Editor's Note: California scientists using recombinant DNA techniques have created a synthetic gene that can be used to make human insulin in a laboratory. The achievement was announced in September 1978 at a seminar at the University of California at Los Angeles.

The nation's first P-4 laboratory for doing high-risk recombinant DNA experiments opened in the spring of 1978 at Fort Detrick, Md.

GENETIC RESEARCH

M ANY AMERICANS have never heard of recombinant DNA —a gene-splicing technique which enables scientists to combine the genetic material DNA (deoxyribonucleic acid) of different species and create new or drastically altered forms of life. Yet experiments in this relatively new area of genetics could have as great an impact on our lives as the splitting of the atom. "The discovery of recombinant DNA is one of the more striking technological achievements of our century," declared biochemist Liebe F. Cavalieri of the Sloan-Kettering Institute for Cancer Research.[1]

Like atomic energy, recombinant DNA research has the potential for great benefits and grave perils. Some scientists fear that these experiments could create dangerous life forms which, if they escaped from the laboratory, might unleash uncontrollable diseases or alter the course of evolution. Others say the risks are minimal and they claim that this research could revolutionize agriculture, greatly simplify control of pollution and lead to cures for diseases like cancer. Caught in the middle is the citizen who does not know which side to believe.

Adding to the public's confusion is the fact that the concerns over recombinant genetic engineering were raised initially by the very scientists doing the work. The issue came to public attention in July 1974 after a group of prominent scientists proposed a voluntary moratorium on certain gene-splicing experiments until the potential risks could be studied and proper safety measures could be worked out. The moratorium was lifted the following February after a group of scientists met at Asilomar, Calif., and adopted strict guidelines for all future research. The Asilomar guidelines were replaced by a stricter and more detailed set issued in June 1976 by the National Institutes of Health.

The guidelines have not ended the controversy over recombinant DNA. Not all scientists are satisfied that the guidelines provide adequate safeguards against potential hazards. Furthermore, the guidelines apply only to research funded by the National Institutes of Health and other federal agencies. Concern over the adequacy of the guidelines has prompted

---

[1] Liebe F. Cavalieri, "New Strains of Life—or Death," *The New York Times Magazine,* Aug. 22, 1976, p. 8.

several communities, including Cambridge, Mass., to adopt or consider further restrictions on the conditions under which such research may proceed *(see p. 118)*.

At the federal level, a committee representing 16 government agencies issued a report on March 15 urging Congress to require federal licensing of all laboratories doing recombinant DNA research. The House Subcommittee on Health and Environment held hearings March 15-17 to consider a bill introduced by its chairman, Rep. Paul G. Rogers (D Fla.), to achieve that end. Similar bills have been introduced by Rep. Richard L. Ottinger (D N.Y.), Rep. Stephen J. Solarz (D N.Y.) and Sen. Dale Bumpers (D Ark.). The Senate Subcommittee on Health, headed by Sen. Edward M. Kennedy (D Mass.), will consider these bills and other aspects of the recombinant DNA controversy at hearings scheduled for mid-April. "Not since the congressional investigations of atomic energy in the 1950s has science sparked such heated political discussion," observed Arthur Lubow, an associate editor of *New Times* magazine.[2]

Advancing Technology of Recombinant DNA

The development of recombinant DNA technology is considered by many scientists to be the most important advance in molecular biology since the discovery of the structure of DNA in 1953. *(see p. 113)*. The analytic power of the new technology has led some scientists to compare it with the invention of the microscope. "The ability to combine genes of different species in a growth medium is the most powerful tool to come along in my lifetime," declared Dr. Frederick Neidhardt, chairman of the microbiology department at the University of Michigan.[3]

The scientific technique involved is relatively simple to understand. By using a substance called a restriction enzyme, researchers can separate DNA molecules at specific points and then recombine them with DNA segments separated from another source. The resulting hybrids are inserted into bacteria in which they reproduce. In this way, genes wanted for study can be produced in large quantities. Sometimes researchers use viruses (tiny life forms consisting largely of DNA) to carry genes into bacteria or other host cells. Another way to get foreign DNA into bacteria is to use something called a plasmid *(see p. 105)*. This is a small circular piece of DNA found naturally in bacteria; it can move easily from one cell to another.

For the moment the chief value of recombination is that it provides scientists with a way of learning more about how genetic molecules function. While biologists already know a

---

[2] Arthur Lubow, "Playing God With DNA," *New Times,* Jan. 7, 1977, p. 52.
[3] Quoted in *Business Week,* Aug. 9, 1976, p. 66.

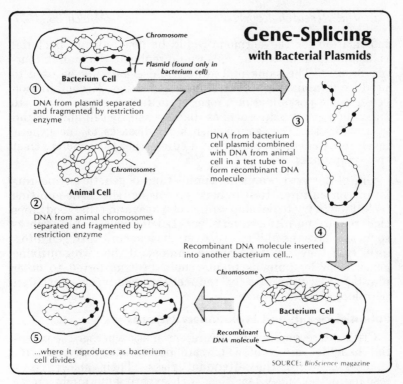

Gene-Splicing

with Bacterial Plasmids

① Bacterium Cell — Chromosome — Plasmid (found only in bacterium cell)

DNA from plasmid separated and fragmented by restriction enzyme

② Animal Cell — Chromosomes

DNA from animal chromosomes separated and fragmented by restriction enzyme

③ DNA from bacterium cell plasmid combined with DNA from animal cell in a test tube to form recombinant DNA molecule

④ Recombinant DNA molecule inserted into another bacterium cell...

Chromosome — Bacterium Cell — Recombinant DNA molecule

⑤ ...where it reproduces as bacterium cell divides

SOURCE: *BioScience* magazine

great deal about the operation of DNA in bacteria, they understand much less about its activity in higher organisms such as human beings. According to Dr. Stanley N. Cohen, a molecular geneticist and professor of medicine at the Stanford University School of Medicine, the use of recombinant DNA technology has already provided scientists with knowledge about how genes are organized into chromosomes and how gene expression is controlled. "With such knowledge," he told the Committee on Environmental Health of the California Medical Association last Nov. 18, "we can begin to learn how defects in the structure of such genes alter their function."[4] The research also could provide scientists with improved understanding of the way in which cells—including cancer cells—reproduce.

In addition to recombinant DNA's potential contributions to the advancement of fundamental scientific and medical knowledge, there are possible practical applications as well. For example, General Electric has applied for a patent on a process that will use the recombinant technique to create bacteria capable of absorbing oil for cleaning up spills. In medicine, it might be possible to "teach" bacteria to produce inexpensive and abundant quantities of human insulin, blood clotting factors, and other valuable hormones. DNA containing the genetic

---

[4] Stanley N. Cohen, "Recombinant DNA—Fact and Fiction," *Science*, Feb. 18, 1977, p. 655.

information for the hormone would be inserted into bacterial cells and, it is hoped, the cells would replicate the manufacturing process. The same procedure could eventually be used to produce vitamins, antibiotics and other drugs. Recombination could make possible a new form of medicine, gene therapy, to treat such genetic disorders as diabetes, sickle cell anemia and cystic fibrosis. By dealing with such diseases at the genetic level, researchers hope to effect a cure rather than merely treat the symptoms.

One of the most promising applications of genetic engineering is in agriculture. Researchers are using the gene-splitting technique to try to develop strains of wheat and other food crops that require no nitrogen fertilizer. Leguminous plants such as soybeans have the ability to convert atmospheric nitrogen into a form that they can use for nourishment. If the nitrogen-fixing genes from leguminous plants could be transferred to other plants, it would greatly reduce the need for synthetic, petroleum-based fertilizers.

Potential Health and Evolutionary Hazards

Critics of recombinant DNA research ask whether the potential benefits—many of which are purely speculative at this time—are worth the attendant risks. Their answer is a resounding no. They fear that such experiments might create Andromeda-type germs which could unleash uncontrollable diseases. For example, it is said that disease-producing bacteria like streptococci could, as a result of genetic engineering, accidently be made immune to antibiotics and other drugs used to treat them. Similarly, a bacterium that now inhabits the human body without doing harm might receive a genetic transplant that would cause it to begin manufacturing a deadly toxin. Moreover, it could spread undetected for a long time.

Harold M. Schmeck, a science reporter for *The New York Times,* wonders what would happen if some of the anticipated benefits backfired. "What if the postulated oil-gobbling bacteria got loose and became a contagious disease of automobiles, aircraft and all other machinery lubricated by oil?" he asks. "What if the insulin-producing bacteria learned to thrive inside humans and somehow sent every infected person into insulin shock? What if scientists inadvertently produced a super germ or a super weed capable of upsetting the entire balance of life on earth?"[5]

Most discussions of risks center on potential health hazards. But there is also concern about the long-range effect on the master plan of evolution. The leading spokesman for the prophets of evolutionary disaster is Dr. Robert L. Sinsheimer,

[5] *The New York Times,* Feb. 20, 1977.

Human Cloning—Next Step?

Some people maintain that current work in recombinant DNA is a first step in the direction of human cloning, or the multiplication of large numbers of genetically identical individuals. The process already is being used with other species. It has been used successfully with plants, fruit flies, and, more recently, with frogs.

Although the cloning of human beings is not yet technically feasible, a few researchers say it would be possible and desirable in the near future. Writing about cloning in *Psychology Today* in June 1974, Dr. James D. Watson said that if such technology "proceeds in its current nondirected fashion, a human being born of clonal reproduction most likely will appear on earth within the next 20 to 50 years, and even sooner if some nation actively promotes the venture."

Although most scientists—including most of those involved in recombinant DNA research—oppose the idea of human cloning, a few defend it. Dr. Joshua Lederberg of Stanford University has said: "If a superior individual is identified, why not copy it directly...?"*

*Quoted in *The Human Agenda* (1972), by Roderick Gorney, M.D., p. 222.

chairman of the biology division at the California Institute of Technology. Sinsheimer maintains that there are natural barriers to genetic interchange between cells of higher organisms (such as man) and cells of lower organisms (such as bacteria). To break down these natural barriers is to risk causing unpredictable damage to the evolutionary process. "The point is that we will be perturbing, in a major way, an extremely intricate ecological interaction which we understand only dimly," he said.[6]

Sinsheimer considers the deliberate misuse of gene-splicing a serious possibility. The problem is analogous to that of nuclear terrorism, he said. "It may well be that there are some technologies that you should not use not because they can't work but because of the social dangers involved and the repression that would be necessary to prevent social danger."[7] The 1972 treaty on biological warfare signed by the United States and 110 other nations would cover toxins or other biological agents developed through recombinant DNA research. But the Federation of American Scientists observed that "treaties are neither universal nor self-enforcing," and therefore "the world must begin to face a biological proliferation threat that might, before long, rival that of nuclear weapons."[8]

[6] Quoted by William Bennett and Joel Gurin in "Science that Frightens Scientists; The Great Debate Over DNA," *The Atlantic,* February 1977, p. 58.
[7] Quoted by Nicholas Wade in "Recombinant DNA: A Critic Questions the Right to Free Inquiry," *Science,* Oct. 15, 1976, p. 305.
[8] *F.A.S. Public Interest Report,* April 1976, p. 1.

Of special concern is the relative simplicity of the experiments. Science writer Judith Randal recounted the story of a Massachusetts Institute of Technology undergraduate who, after having read published reports, demonstrated on paper that he knew how to build an atomic bomb. "Since recombinant DNA work requires only a meager investment in equipment and can be carried out in a limited space," she wrote, "a similarly resourceful high school student could conceivably collect the necessary materials and then simply turn the experimental brew loose on the general environment."[9]

Those who support recombinant DNA research emphasize that scientists know of no hazardous agent it has ever created. Critics concede this point, but insist that until the potential risks can be accurately determined and assessed, the research should be restricted, postponed or banned altogether. Proponents find this approach unsatisfactory. They argue that no one will ever be able to guarantee total freedom from risk in any significant human activity. "All that we can reasonably expect," Dr. Cohen has said, "is a mechanism for dealing responsibly with hazards that are known to exist or which appear likely on the basis of information that is known."

Lab Rules for Genetic Altering of Microbes

At present, the principal mechanism for balancing the potential benefits and risks of genetic engineering is the set of guidelines issued by the National Institutes of Health on June 23, 1976. The guidelines provide two lines of defense against the escape of genetically altered microbes—physical and biological. There are four levels of physical containment:

P1 *(minimal):* strict adherence to standard practices.

P2 *(low):* limited access to laboratory during experiments; precautions against the release of aerosols and the prohibition of mouth pipetting.

P3 *(moderate):* laboratories equipped to ensure inward air flow; requires use of safety cabinets, the wearing of gloves by personnel, and decontamination of recirculated air.

P4 *(high):* special facilities of the kind used in biological warfare testing; requires rooms equipped with air locks, clothing changes and showers before leaving work area, and decontamination of all air, liquid and solid wastes.

Biological containment calls for the use of enfeebled strains of bacteria, as experimental hosts, which supposedly cannot survive outside the laboratory. A complex set of rules specifies which types of experiments require what combinations of physical and biological containment. Some experiments judged

[9] Judith Randal, "Life From the Labs: Who Will Control the New Technology?" *The Progressive,* March 1977, p. 18. Randal is a science writer for the *New York Daily News.*

especially dangerous are banned altogether. These include increasing the virulence of known pathogens or making microbes more resistant to antibiotics. No work is allowed with genetic material from organisms that produce dangerous poisons, such as botulism or snake or insect venom. No recombinant DNA molecule may be released into the environment, no matter how benign it is believed to be.

Most scientists seem satisfied with the guidelines, to judge from a straw poll conducted by the Federation of American Scientists last year.[10] The results showed that 56 per cent of the federation's members, including 64 per cent of the biologists, thought the guidelines were "probably about right." However, 7 per cent, including 12 per cent of the biologists, considered the guidelines to be overly restrictive. They regarded the potential dangers as wholly speculative and exaggerated; they were more concerned about the restraints on freedom of inquiry posed by the guidelines.

On the other hand, a substantial number of federation members—29 per cent, including 32 per cent of the non-biologists—said the guidelines did not go far enough. "The guidelines may alleviate the nervousness of some scientists but...my own view is that they will not effectively reduce the danger," said Liebe Cavalieri. "Indeed they may actually lull us into a false sense of security." Cavalieri said that in the course of 25 years of Army research with biological warfare agents at Fort Detrick, Md., equipped with the highest level of physical containment facilities, there were 423 accidental infections and three deaths.

One federal agency with serious reservations about the guidelines is the Environmental Protection Agency. The agency is particularly concerned that the guidelines permit researchers to continue using a type of bacteria that is commonly found in the human intestine, E. Coli *(Escherichia coli)*, as a host for recombinant DNA molecules. Wilson K. Talley, the agency's research director, told the Senate health subcommittee on Sept. 22, 1976, that such research "should be performed on organisms...which are less ubiquitous than E. Coli." Voicing a similar concern, Dr. Erwin Chargaff of Columbia University said in a letter published in *Science* magazine last July 4: "If Dr. Frankenstein must go on producing his little biological monsters...why pick on E. Coli as the womb? Why choose a microbe that has cohabited more or less happily with us for a long time indeed?"

Other scientists dismiss such concerns. They insist that the wealth of existing knowledge about E. Coli and its genetic

---

[10] The poll was taken in April 1976 and the results made public on June 23, 1976.

makeup (E. Coli is the traditional laboratory bacterium) will make it a safer host than any other bacterium. They point out that the guidelines require the use of enfeebled strains of E. Coli that are not typical residents of the human intestine. But critics of recombinant research fear that even crippled forms of this microbe might transmit dangerous characteristics to ordinary E. Coli if they were accidently released.

Still other scientists, led by Dr. Sinsheimer, criticize the guidelines as being narrowly concerned with safety and not addressing broad moral, ethical or evolutionary questions raised by this research. In their opinion, there was not enough public participation in the process that led to publication of the guidelines. They say that those who had the greatest voice in formulating the guidelines were scientists already committed to going ahead with the research.

Industry's Freedom From U.S. Guidelines

A key problem with the NIH guidelines is enforcement. Although other federal agencies have adopted them, they do not cover private industries or laboratories with independent funding and they do not have the force of law. "As of now, there is no federal agency that is looking at research being done by private industry in recombinant DNA research," Dr. Bernard Talbot of NIH said in an interview published in *New Times* magazine Jan. 14. "[I]f private industry wants to do research, the federal government has no right to inspect or monitor the facilities...."

According to the Peoples Business Commission (formerly the Peoples Bicentennial Commission)[11] seven drug companies are engaged in or are about to begin recombinant DNA research: Hoffman-LaRoche in Nutley, N.J.; Eli Lilly in Indianapolis, Ind.; Upjohn in Kalamazoo, Mich.; Miles Laboratories in Rochester, N.Y., and South Bend, Ind.; Merck, Sharp & Dohme Research Laboratories in Rahway, N.J.; Abbot Laboratories in North Chicago, Ill.; and Pfizer in Groton, Conn. Nine other companies, the Peoples Business Commission said, are now looking into the potential applications of recombinant DNA. They are Cetus, CIBA-Geigy, DuPont, Dow, W. R. Grace, Monsanto, French Laboratories, Wyeth Laboratories and Searle Laboratories.[12] *Medical World News* reported Oct. 4 that "probably no one person in America has any clear idea of just how much recombinant research is already going on in industry or with what precautions."

---

[11] The Peoples Business Commission is a nonprofit, educational organization established to increase public awareness of corporate power and policies.
[12] See Jeremy Rifkin, "DNA," *Mother Jones*, February-March 1977, p. 23. *Mother Jones* is a national monthly magazine of news, politics, commentary and the arts published by the Foundation for National Progress in San Francisco.

Patents for New Life Forms

One problem with the current NIH guidelines on recombinant DNA is that they do not apply to private industries or independently funded laboratories engaged in such research. The Pharmaceutical Manufacturers Association, which represents the major drug companies working with recombinant DNA, has said that the pharmaceutical industry could accept the NIH guidelines with some minor modifications.

One of the drug industry's objections to the guidelines is that they require advance disclosure of research plans. That provision is considered crucial by most scientists. But pharmaceutical companies argue that it interferes with their patent rights.

The Department of Commerce announced on Jan. 10 that firms applying for patents on recombinant DNA techniques would be exempted from the advance disclosure provision. Betsy Ancker-Johnson, Assistant Secretary of Commerce for Science and Technology, said that "in view of the exceptional importance of recombinant DNA and the desirability of prompt disclosure of developments in the field," she also had recommended that the department speed the processing of patent applications that involve gene-splicing.

The department's action was widely criticized. Sen. Dale Bumpers (D Ark.) told his Senate colleagues that recombinant DNA is an area "which is entirely too dangerous to worry about proprietary information." At the request of HEW Secretary Califano, Secretary of Commerce Juanita M. Kreps on Feb. 24 agreed to suspend the order.

Concern over industry's freedom from regulation led Sens. Kennedy and Jacob K. Javits (R N.Y.) to write a letter to President Ford, dated July 19, 1976, urging him to make all recombinant research, including that being conducted by industry, subject to federal control. Responding to the letter, Ford announced Sept. 22 the establishment of an interagency commission headed by Donald Frederickson, director of the National Institutes of Health, "to review the activities of all government agencies conducting or supporting recombinant DNA research or having regulatory authority relevant to this scientific field."

Five months later, on Feb. 19, 1977, Frederickson announced that the commission had concluded that federal legislation did not completely cover the regulation of recombinant research. For example, the Toxic Substances Control Act of 1976 tightened federal regulation of all chemicals and chemical combinations, but it exempted research laboratories. In a report sent to the Secretary of Health, Education and Welfare, Joseph A. Califano Jr., on March 15, the interagency commission recommended that Congress extend federal control to all laboratories doing DNA research. The commission

111

recommended that a new law be enacted to (1) require registry, licensing and inspection of all laboratories where such research is conducted, (2) supersede any local rules, and (3) provide a mechanism to let private firms keep some work secret until they apply for patents *(see box, p. 111)*.

While some members of the scientific community continue to oppose new legislation, most scientists attending a meeting of the National Academy of Sciences in Washington, D.C., March 7-9, appeared to support such a move. Expressing the view of many of his colleagues, Dr. Daniel Koshland, chairman of the biochemistry department of the University of California at Berkeley, said: "If there is no federal legislation then every city will make its own rules."

Some scientists and groups, including the Peoples Business Commission, have urged a more drastic step—an immediate moratorium on all recombinant DNA research. To push for such a moratorium, three scientists attending the Washington meeting—Dr. Ethan Signer of the Massachusetts Institute of Technology and Drs. Jonathan Beckwith and George Wald of Harvard—announced the formation of an international Coalition for Responsible Genetic Research. They said: "The continuation of this research without public understanding and approval and, in fact, without a full comprehension of its potential by most of the involved scientists, poses a worldwide danger which is intensified by the fact that industrial investment in the developing genetic technology has already begun."

Rise of Concern Among Scientists

T HE STUDY of genetics has come a long way since Gregor Mendel's experiments with pea plants demonstrated the fundamental laws of inheritance. Mendel, a 19th-century Austrian monk, showed that for each physical trait, every individual possessed two "factors," or what later came to be known as genes.[13] Biologists had only fragmentary knowledge of the genetic process until the mid-20th century. It was known that genes were arranged in linear sequence along chromosomes, which are present in the nucleus of every living cell, but nothing was known about the molecular structure of genetic material.

In 1944, however, three biochemists at the Rockefeller Institute—Oswald T. Avery, Colin MacCloud and Maclyn McCarty—learned that genes were composed of deoxy-

---

[13] Mendel's findings were published in 1866 in an obscure journal, *The Proceedings of the Natural History Society of Bruenn,* and aroused little interest until 1900.

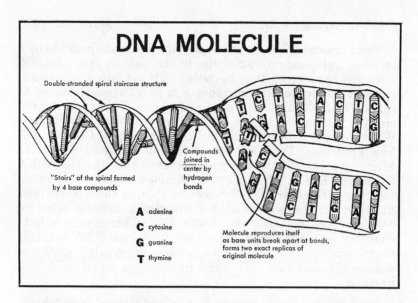

DNA MOLECULE

Double-stranded spiral staircase structure

"Stairs" of the spiral formed
by 4 base compounds

Compounds
joined in
center by
hydrogen
bonds

A adenine

C cytosine

G guanine

T thymine

Molecule reproduces itself
as base units break apart at bonds,
forms two exact replicas of
original molecule

ribonucleic acid (DNA). The next step was to determine the structure of the complex DNA molecule. This was accomplished in England in 1953 by two Cambridge scientists, Francis H. C. Crick and Maurice H. F. Wilkins, and an American colleague, James D. Watson.[14] The three scientists, employing X-ray diffraction pictures of DNA crystals, chemical analyses and their own intuition, concluded that the DNA molecule had a double-helix or spiral-staircase structure. According to their calculations, the main strains forming the backbone of the molecule are composed of long, spiral chains of sugar-phosphate units, endlessly repeated. These strands are joined together at regular intervals by small side chains, or "base" units, to complete the spiral-staircase structure.

The base units, each attached at one side to a strand unit and joined in the center by a hydrogen bond, are generally composed of one of four related compounds: adenine (A), thymine (T), guanine (G), or cytosine (C). Compound A will join chemically only with T, and G only with C. From this information, biologists deduced the method by which a DNA molecule reproduces itself. First the two main strands separate down the middle, forming a pair of templates or molds.

Thus, for instance, if it could be determined that a portion of one strand [of a DNA molecule] contains the [base] sequence AGGTCGCAT, then it would follow that the second strand would contain, at the corresponding site, the sequence TCCAGCGTA. The relationship between the two would be that between a photographic negative and the corresponding positive.[15]

---

[14] See Watson's account of their discovery in his book *The Double Helix* (1968).

[15] Isaac Asimov, "The Genetic Code," *New York State Journal of Medicine*, June 15, 1965, p. 1648. See also "Genetics and the Life Processes," *E.R.R.*, 1967 Vol. II, pp. 903-922.

Recombinant DNA, or gene-splicing, was made possible by a series of independent discoveries in the past 15 years. In 1962 scientists discovered that bacterial cells contain a substance, called a restriction enzyme, that acts as a chemical scalpel to split DNA molecules into specific segments. Ten years later the enzyme was purified from bacteria by microbiologist Herbert W. Boyer and his colleagues at the University of California at San Francisco. Then two researchers at Stanford—Janet E. Mertz and Ronald W. Davis—discovered that the split DNA fragments had sticky ends that enabled them to be joined together with other DNA fragments. Putting these discoveries together, Dr. Stanley N. Cohen of Stanford University School of Medicine and his assistant, Annie C. Y. Chang, were able to construct in a test tube a biologically functional DNA molecule that combined genetic information from two different sources—in this case, two different plasmids found in E. Coli bacteria.

Subsequent experiments by Cohen and Chang, in collaboration with Herbert W. Boyer and Robert B. Hellig of the University of California at San Francisco, showed that genes from another species of bacterium, Staphylococcus aureus, could be transplanted into E. Coli. Further experimentation demonstrated that animal DNA—specifically, ribosomal DNA from the South African toad—could be linked with plasmid DNA to form recombinant molecules that would reproduce in E. Coli. A proliferation of recombinant DNA work followed, resulting in the insertion into bacteria of animal DNA from fruit flies, toads, mice, sea urchins, slime molds and chickens.[16]

Organized Effort to Assess Research Risks

Cohen and his colleagues recognized from the beginning that the construction of some kinds of novel gene combinations might have a potential for biological hazard. At first the primary concern was that certain gene-splicing experiments might increase the risks of work with cancer viruses. One of the first scientists to voice this concern publicly was Paul Berg of Stanford, who decided to abandon plans to introduce genes from a tumor virus into E. Coli bacteria after his colleagues suggested that the resulting organism might spread cancer to humans.

Berg helped to organize a conference in New Hampshire in July 1973 to review the available information on recombinant research and to assess the potential risks. Those attending the conference—the Gordon Research Conference on Nucleic Acids—sent an open letter to Dr. Philip Handler, president of the National Academy of Sciences, warning that new organisms

---

[16] See Stanley N. Cohen, "The Manipulation of Genes," *Scientific American*, July 1975, pp. 24-33.

"with biological activity of an unpredictable nature" could be created by these experiments. They urged him to "establish a study committee to consider this problem and to recommend specific actions or guidelines should that seem appropriate."[17]

A committee was formed, with Paul Berg as chairman, and in a now-famous letter to *Science* magazine in July 1974, it recommended that certain types of recombinant DNA research be voluntarily deferred "until the potential hazards...have been better evaluated or until adequate methods are developed for preventing their spread...."[18] The types of research covered by the moratorium were (1) formation of bacteria resistant to antibiotics, (2) linkage of DNA molecules with tumor-causing viruses, and (3) introduction of toxin-formation or antibiotic-resistance genes into bacteria that did not naturally contain such genes. Berg's committee also asked the director of the National Institutes of Health, Robert S. Stone, to set up an advisory committee to evaluate potential hazards in this research, devise safety procedures and develop guidelines for researchers working with potentially hazardous DNA molecules. Finally, the Berg committee said that an international conference should be convened as soon as possible.

The Berg committee's call for a voluntary moratorium was called an unprecedented event. In fact, the American Chemical Society listed it among the most important scientific events of the last 100 years.[19] The moratorium was widely reported by the press and produced an international reaction. Stone quickly announced his intention to establish an advisory committee, as recommended, and he offered financial support for an international meeting. In England, the Advisory Board for the Research Councils, the prime source of government funding for civil research in Britain, set up a committee to assess the potential hazards and benefits of genetic engineering. In the interim, the board asked all of its units to suspend any experiments cited by the Berg committee as particularly dangerous.

In his presidential address to the British Association for the Advancement of Science in September 1974, molecular biologist Sir John Kendrew commended the Berg committee's actions and suggested the establishment of a permanent international monitoring body of molecular biologists who would assess gene-transfer experiments. On the other hand, the influential British journal *Nature,* in an editorial on Sept. 6, 1974, rejected a suggestion that it cease publication of articles on research

---

[17] The letter was published in *Science,* Sept. 23, 1973, p. 1114.

[18] *Science,* July 26, 1974, p. 303.

[19] *Chemical & Engineering News,* April 6, 1976. *Chemical & Engineering News* is the official publication of the American Chemical Society.

covered by the proposed moratorium. In October 1974, DNA recombination came under discussion at a Pugwash Conference in Austria and at an international symposium in Davos, Switzerland. Most participants at the Davos conference acknowledged the "enormous dangers" posed by recombination, but they concluded that controls would be "impractical and unenforceable."[20]

Recommendations From 1975 Conference

The international meeting proposed by the Berg committee was held in February 1975 at the Asilomar Conference Center in Pacific Grove, Calif. It was sponsored by the National Academy of Sciences and was supported by the National Institutes of Health and the National Science Foundation; 150 persons from 16 countries attended. They revealed a wide divergence of opinion in the scientific community. Nobel laureate Joshua Lederberg of Stanford expressed his dismay at the prospect that guidelines might end up "crystallized into legislation." James D. Watson of Harvard said that guidelines would be essentially unenforceable, and that therefore the best tactic was to rely on the common sense of those doing the research. Some participants argued that the risks were too remote to justify limiting the freedom of scientific inquiry. Others insisted that the moral responsibility to protect the public was more important than academic freedom or individual success.

In the end, the participants concluded that "most of the work...should proceed." They ranked the experiments by potential risk and specified safety precautions for each level. And they favored a ban on experiments that, while feasible, "present such serious dangers that their performance should not be undertaken at this time."[21]

Nicholas Wade of *Science* magazine described the actions taken at Asilomar as "a rare, if not unique, example of safety precautions being imposed on a technical development before, instead of after, the first occurrence of the hazard being guarded against."[22] Jack McWethy of *U.S. News & World Report* called the conference a landmark "because it provided for all scientists a working illustration of how specialists can examine and, when necessary, limit their research for the public good long before the issues are dragged into the...political arena."[23]

---

[20] See *Science News,* Nov. 2, 1974, p. 277, and *BioScience,* December 1974, p. 694.

[21] A report on the conference was submitted to the Assembly of Life Sciences of the National Academy of Sciences and was approved by its executive committee, May 20, 1975. See Janet H. Weinberg, "Decision at Asilomar," *Science News,* March 22, 1975, pp. 194-196, and Cristine Russell, "Biologists Draft Genetic Research Guidelines," *BioScience,* April 1975, pp. 237-240.

[22] Nicholas Wade, "Genetics: Conference Sets Strict Controls to Replace Moratorium" *Science,* March 14, 1975, p. 931.

[23] Jack McWethy, "A Move to Protect Mankind," *U.S. News & World Report,* April 7, 1975, p. 66.

Immediately after the conference, the NIH Advisory Committee on Recombinant DNA, which had been set up in October 1974, held its first meeting in San Francisco to begin translating the mandate of Asilomar into firm guidelines binding on all researchers receiving NIH grants. At a second meeting, held May 12-13, 1975, in Bethesda, Md., a subcommittee under the chairmanship of Dr. David Hogness was appointed to draft the guidelines. The first draft, made public the following July 18-19 at a meeting in Woods Hole, Mass., was widely criticized as being weaker than the rules agreed upon at Asilomar. Two Boston-centered groups, Science for the People and the Boston Area Recombinant DNA Group, organized a petition drive against the draft guidelines. Eventually a new NIH subcommittee was appointed to revise them.

The draft guidelines finally adopted by the NIH advisory committee at La Jolla, Calif., on Dec. 5, 1975, were, according to Nicholas Wade, "demonstratively stricter than the Asilomar guidelines...."[24] Although the guidelines were criticized in some quarters, they were approved by the National Institutes of Health and released on June 23, 1976.

During the months of debate that preceded the issuing of the guidelines, questions were raised about the possibility that organisms containing recombinant DNA molecules might escape and harm the environment. The National Institutes of Health pointed out that the guidelines prohibit the deliberate release of such organisms. Nevertheless it agreed to review the possible environmental impact of genetic experiments. A draft of the environmental impact statement was released for public comment on Aug. 26, 1976.[25]

Emergence of Citizen Oversight

SCIENCE TODAY is facing the equivalent of the Protestant Reformation, according to University of Chicago philosopher Stephen Toulmin. Likening the scientific establishment to the 16th century church, Toulmin said that the people are tired of being shut out of science's "ecclesiastical courts" and are demanding to be let in. The scientist "priest," he predicted, is going to be overthrown.[26]

---

[24] Nicholas Wade, "Recombinant DNA: NIH Sets Strict Rules to Launch New Technology," *Science*, Dec. 19, 1975, p. 1175.

[25] See the *Federal Register*, Sept. 9, 1976, pp. 38426-38483. The NIH guidelines were published in the *Federal Register*, July 7, 1976, pp. 27902-27943.

[26] Quoted in Barbara J. Culliton, "Public Participation in Science; Still in Need of Definition," *Science*, April 30, 1976, p. 451.

In the past, the public tended to acquiesce in the judgment of scientists in the assumption that any advance of knowledge was necessarily beneficial. But in recent years, trust and approval have given way to suspicion and apprehension among increasing numbers of Americans. Behind the public's misgivings is a litany of known or suspected hazards that were the product of scientific research: DDT, cyclamates, asbestos, PCB, vinyl chloride, radioactivity, aerosol propellants, food additives, Kepone.

The upshot of all this appears to be that "the American public is coming to regard scientific research with what might be termed a *Code Napoleon* attitude," according to George Alexander, science writer for the *Los Angeles Times.* "Just as that French legal system presumes that an individual is guilty of an alleged crime and places the burden of innocence upon the accused, so does this evolving public attitude presuppose that new research is more likely to be harmful than beneficial, that disadvantages are more likely to outweigh advantages."[27]

Nowhere is this public attitude more evident today than in the debate over the safety of recombinant DNA research. In barely five years, it already has given rise to Vietnam-type protest groups and to city council and state legislative hearings from Cambridge, Mass., to Sacramento, Calif. "Gene transplantation may be the first innovation submitted to public judgment *before* the technology had been put into widespread use and before heavy investment had given it a momentum that was hard to oppose."[28]

The first local rumblings of discontent came in Ann Arbor, Mich. Early in 1975, the regents of the University of Michigan began to consider a plan to upgrade some laboratories to the P3 level of containment *(see p. 108).* This set off a debate which went on for over a year. At several public hearings the plan was opposed by the Ann Arbor Ecology Center and a few faculty members, notably Shaw Livermore, a professor of American history, and Susan Wright, associate professor of humanities. Despite the opposition, the regents in May 1976 voted 6 to 1 to proceed with the research.

Community Action on Recombinant Research

The debate in Ann Arbor was largely confined to the university. A much broader public debate took place in Cambridge, Mass. The controversy over recombinant DNA erupted in June 1976 after a weekly newspaper, *The Boston Phoenix,* reported Harvard's plan to convert an existing laboratory into a P3

---

[27] *Los Angeles Times,* Feb. 27, 1977.
[28] Bennett and Gurin, *op. cit.,* p. 44.

facility. Some faculty members, led by Nobel laureate George Wald and his wife, Harvard biologist Ruth Hubbard, expressed their opposition to Cambridge Mayor Alfred E. Vellucci. He called a public meeting on the matter, saying "We want to be damned sure the people of Cambridge won't be affected by anything that would crawl out of that laboratory."

The Cambridge City Council considered the issue on June 23 at a hearing attended by nearly 500 persons and again on July 7. At the second meeting the council imposed a three-month moratorium on moderate and high-risk DNA experiments until a citizens' review board could study the problem. This was considered a precedent-setting action for involving the public in decision-making regarding biological research.[29]

The nine members of the Cambridge Experimentation Review Board met twice a week for five months (the moratorium was extended) and issued a report in January 1977 declaring that "knowledge, whether for its own sake or for its potential benefits to mankind, cannot serve as a justification for introducing risks to the public unless an informed citizenry is willing to accept those risks." The review board decided unanimously that it was prepared to accept those risks, and it recommended that the research be allowed to continue.

However, in the belief that "a predominantly lay citizen group can face a technical scientific matter of general and deep public concern, educate itself appropriately to the task, and reach a fair decision," the panel concluded that the safety guidelines developed by the National Institutes of Health did not go far enough. The board recommended some additional measures, including the preparation of a safety manual, training of personnel to minimize accidents, and inclusion of a community representative on the NIH-mandated "biohazards committees" at Harvard and the Massachusetts Institute of Technology. The review board also recommended that the city set up a permanent citizen biohazards committee to monitor the research at the universities and report violations.

On Feb. 7, 1977, the Cambridge City Council endorsed the board's recommendations after rejecting a proposal by Mayor Vellucci to ban the research altogether. "What happened in Cambridge is of major national importance," said Stanley Jones, a staff member of the Senate Health Subcommittee. "It is the first time a public community group has looked at an issue in science and made recommendations on what it thought was appropriate."[30]

---

[29] See Nicholas Wade, "Recombinant DNA: Cambridge City Council Votes Moratorium," *Science,* July 23, 1976, p. 300.

[30] Quoted in *The Christian Science Monitor,* Jan. 17, 1977.

The year-long debate in Cambridge spurred action in other communities. In San Diego, Calif., the city's Quality of Life Board acted at the request of Mayor Pete Wilson to set up a committee to review DNA work at the University of California at San Diego. After hearing an array of witnesses, the committee in February 1977 submitted a report generally endorsing the NIH guidelines. But in addition, it recommended that (1) the city council consider the desirability of confining all gene-splicing research to P3 laboratories, (2) the university refrain from experiments requiring P4 facilities, (3) it notify the city of any P3 experiment requiring the highest degree of biological containment, and (4) an ordinance be passed to bring industry and private researchers within the control of the guidelines.

In Madison, Wis., the city council recently appointed a committee to study the possible hazards of recombinant DNA research at the University of Wisconsin. A citizen review board also was set up recently in Princeton, N.J. Public hearings on the question have been held in several other university towns, including Bloomington, Ind. (Indiana University), New Haven, Conn. (Yale), and Palo Alto, Calif. (Stanford). At the state level, bills to control DNA recombination experiments have been introduced in New York and California."What all these activities represent..." Nicholas Wade wrote, "is an extended exercise in public education about the gene-splice technique and its implications.... Whatever further restrictions emerge from the present round of debate, the research will at least be proceeding on the basis of informed public consent...."[31]

Significance of Public Participation in Science

The calls to prohibit or slow down the research seem threatening and irrational to many of those scientists who first pointed to the risks. The public's response might make scientists reluctant to question the consequences of any future research for fear of generating "unjustified fears" and "opening themselves up to attack," according to Professor Mark Ptashne of Harvard.[32] Dr. Cohen writes that the public has misinterpreted the scientific community's attempts at self-regulation "as *prima facie* evidence that this research must be more dangerous than all the rest."

Because in the past, governmental agencies have often been slow to respond to clear and definite dangers in other areas of technology, it has been inconceivable to scientists working in other fields and to the public at large that an extensive and costly federal machinery would have been established to provide protec-

---

[31] Nicholas Wade, "Gene-Splicing: At Grass-Roots Level a Hundred Flowers Bloom," *Science*, Feb. 11, 1977, p. 560.
[32] Quoted in *The Chronicle of Higher Education*, Aug. 2, 1976, p. 4.

Regulation of Recombinant DNA Abroad

No matter how tough American regulations on recombinant DNA become, there is little the United States can do to control gene-splicing experiments in other countries. Some scientists have expressed concern that companies bent on avoiding federal control might set up shop in nations without strict regulations—much as oil-tanker companies use Liberia as a convenient place to register their ships. "Any accident on this globe would affect some other place," warned Dr. J. E. Rall, a research director at the National Institutes of Health.*

The only comprehensive guidelines for recombinant-DNA research outside of the United States are in Britain. The British guidelines differ from America's in that they place more emphasis on physical containment and less on biological containment. The Soviet Union is reported to be drawing up its guidelines based on the U.S. and British models.

In addition there are national committees for genetic engineering in France, Germany, Italy, Belgium, the Netherlands and the Scandinavian countries. Some of these national committees have governmental status and the authority to establish guidelines, inspect laboratories, authorize experiments, and so on; others are merely advisory.

Last October, the International Council of Scientific Unions (ICSU) voted to set up a committee to monitor research associated with recombinant-DNA and other experiments in genetic manipulation. The new body, known as the Committee on Genetic Experimentation (COGE), will conduct no research of its own. But it will attempt to monitor experiments in progress throughout the world and serve as a channel of communication among the scientific communities engaged in such research. In addition, the World Health organization has set up a committee to consider health implications of recombinant DNA.

* Quoted in *The Christian Science Monitor,* Feb. 24, 1977.

tion in this area of research unless severe hazards were known to exist.[33]

So far, the scientist's fear of citizen review seems unjustified. Most public bodies that have considered the recombinant DNA controversy have endorsed the guidelines issued by the National Institutes of Health with minor changes. The report submitted by the Cambridge Experimentation Review Board was praised by both critics and supporters of recombination. "It proved that even complex scientific issues can be understood by lay people who devote the necessary time and energy to the problem," wrote Dr. David Baltimore of the Massachusetts Institute of Technology.[34] As science increases its powers to modify all aspects of society, more and more people are taking the time to question the implications of scientific research.

[33] Stanley N. Cohen, "Recombinant DNA: Fact and Fiction," *Science,* Feb. 18, 1977, pp. 656-657.
[34] David Baltimore, "The Gene Engineers," *T.V. Guide,* March 12, 1977, p. 30.

Selected Bibliography

Books

Handler, Philip, ed., *Biology and the Future of Man,* Oxford University Press, 1970.

Watson, James D., *The Double Helix,* Atheneum, 1968.

Winchester, A. M., *Heredity: An Introduction to Genetics,* Barnes & Noble, 1961.

Articles

Bennett, William and Joel Gurin, "Home Rule and the Gene," *Harvard Magazine,* October 1976.

——"Science That Frightens Scientists: The Great Debate Over DNA," *The Atlantic,* February 1977.

Cavalieri, Liebe F., "New Strains of Life—or Death," *New York Times Magazine,* Aug. 22, 1976.

Cohen, Stanley N., "The Manipulation of Genes," *Scientific American,* July 1975.

——"Recombinant DNA: Fact and Fiction," *Science,* Feb. 18, 1977.

Crossland, Janice, "Hands on the Code," *Environment,* September 1976.

Fields, Cheryl, "Can Scientists Be Trusted on Hazardous Research?" *The Chronicle of Higher Education,* Aug. 2, 1976.

"Fruits of Gene-Juggling: Blessing or Curse?" *Medical World News,* Oct 4, 1976.

Gwynne, Peter, "Caution: Gene Transplants," *Newsweek,* March 21, 1977.

——"Politics and Genes," *Newsweek,* Jan. 12, 1976.

Lubow, Arthur, "Playing God With DNA," *New Times,* Jan. 7, 1977.

"Pandora's Box of Genes," *The Economist,* March 5, 1977.

Randal, Judith, "Life from the Labs: Who Will Control the New Technology?" *The Progressive,* March 1977.

Rifkin, Jeremy, "DNA," *Mother Jones,* February-March 1977.

Russell, Cristine, "Weighing the Hazards of Genetic Research: A Pioneering Case Study," *BioScience,* December 1974.

——"Biologists Draft Genetic Research Guidelines," *BioScience,* April 1975.

Science,, selected issues.

Weinberg, Janet H., "Decision at Asilomar," *Science News,* March 22, 1975.

Reports and Studies

Editorial Research Reports, "Genetics and the Life Process," 1967 Vol. II, p. 903; "Human Engineering," 1971 Vol. I, p. 367; "Medical Ethics," 1972 Vol. I, p. 461.

National Institutes of Health, "Recombinant DNA Research Guidelines," *Federal Register,* July 7, 1976.

——"Recombinant DNA Research Guidelines: Draft Environmental Impact Statement," *Federal Register,* Sept. 9, 1976.

U.S. Congress, Senate Subcommittee on Health, "Oversight Hearing on Implementation of NIH Guidelines Governing Recombinant DNA Research," Sept. 22, 1976.

EARTHQUAKE FORECASTING

by

Sandra Stencel

July 16
1 9 7 6

EARTHQUAKE FORECASTING

I N THE EARLY evening of Feb. 4, 1975, Liaoning Province in northeast China was rocked by a major earthquake which destroyed the town of Haicheng. More than a million people lived near the quake's epicenter and nearly 90 per cent of the houses collapsed, yet according to the Chinese there were few casualties. Exactly one year later, on Feb. 4, 1976, an earthquake of similar magnitude[1] struck Guatemala. More than 22,000 persons were killed and 75,000 were injuried.

It was neither luck nor allegiance to Chairman Mao that enabled the Chinese to avert the heavy loss of life suffered by the Guatemalans. Casualties from the Liaoning earthquake were low because Chinese scientists predicted it and the government ordered the populace to leave their houses for tent cities and other open-air shelters. Vehicles were removed from garages and farm animals from barns. Emergency squads were mobilized. Less than six hours after the final evacuation order was given, the earthquake struck.

"The Chinese success" in predicting the Liaoning earthquake "signals that the age of earthquake prediction may be upon us," states Dr. Robert M. Hamilton, who heads the Office of Earthquake Studies in the U.S. Interior Department's Geological Survey.[2] Dr. Frank Press of the Massachusetts Institute of Technology called the Chinese achievement "one of the major events in the history of geophysics."[3] Not too many years ago, earthquake forecasting was considered an effort best left to psychics, astrologers and religious prophets. It was only within the past decade that new theories on the mechanism and cause of earthquakes made earthquake prediction a reputable field for scientific study.

Today, although the science still is in its infancy, most seismologists, geophysicists and geologists are confident that the

---

[1] The Chinese earthquake measured 7.3 on the Richter Scale and the Guatemalan earthquake measured 7.5. Named for Dr. Charles Richter, professor of seismology at the California Institute of Technology, the Richter Scale measures the strength of earthquakes. Because the scale is logarithmetic, each higher number represents a tenfold increase in the magnitude of the tremors. Thus a magnitude 7 quake is ten times as strong as one of magnitude 6, while a magnitude 8 quake is 100 times as strong as one of magnitude 6 and releases 100 times as much energy.

[2] Quoted by Thomas Y. Canby in "Can We Predict Quakes," *National Geographic,* June 1976, p. 835.

[3] Quoted by Allen L. Hammond in "Earthquakes: An Evacuation in China, A Warning in California," *Science,* May 7, 1976, p. 538.

time, place and magnitude of at least some earthquakes can be predicted by monitoring precursory changes in the earth's crust. V. E. McKelvey, director of the U.S. Geological Survey, told a conference on earthquake warning and response, held in San Francisco on Nov. 7, 1975, that "although operational earthquake prediction systems are not likely to be deployed for a number of years, the prediction capability is achieving significant progress."

Prediction of Impending California Earthquake

The first big test of earthquake forecasting in the United States could come within the next nine months. Dr. James H. Whitcomb, a geophysicist at the California Institute of Technology, has predicted that a moderately strong earthquake—one measuring between 5.5 and 6.5 on the Richter Scale—will occur in the Los Angeles area before April 1977. A quake of this magnitude would be comparable to the 1971 San Fernando Valley earthquake which killed 64 persons and caused property damage of $550 million.

Whitcomb announced his prediction on April 15, 1976, at a meeting of the American Geophysical Union in Washington, D.C., after studying seismic wave measurements taken along the San Andreas Fault *(see map)*, the dominant earth fissure in California. This method of earthquake forecasting is based on the theory that certain phenomena occur inside the earth's crust along a fault in the months prior to an earthquake and that these changes can be detected by measuring changes in the speed of sound waves passing through the crust.

Although Whitcomb cautioned that this hypothesis required further testing, scientists and government officials are taking his prediction seriously. The area along the San Andreas Fault is subject to earthquakes. Moreover, a mysterious "bulge" in the earth had been discovered there in 1975 by a team of Geological Survey scientists who were searching through back records of local elevations along the San Andreas Fault. A comparison of elevations before and after 1960 revealed that a crustal blister several inches high had built up along a 100-mile stretch of the fault near the city of Palmdale. The discovery was disturbing because such uplifts—the term preferred by earth scientists—have sometimes, but not always, preceded major earthquakes.

Evidence made public by the Geological Survey in May 1976 revealed that the Palmdale bulge is bigger and wider than it was originally believed to be. Scientists still are not entirely sure what the uplift means. It is possible that the bulge is a preliminary stage of mountain building, said Dr. Wayne Thatcher. But he added that "the rapidity with which the

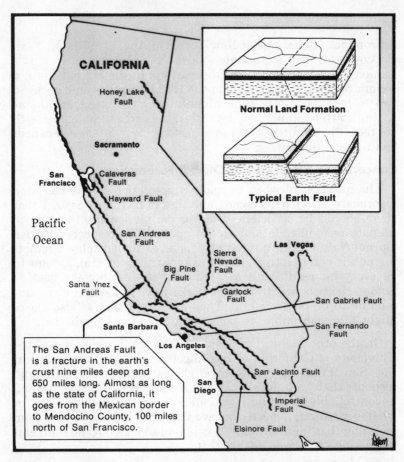

The San Andreas Fault is a fracture in the earth's crust nine miles deep and 650 miles long. Almost as long as the state of California, it goes from the Mexican border to Mendocino County, 100 miles north of San Francisco.

uplift occurred makes that explanation unlikely. It is more likely that the storage of strain will be released someday as an earthquake."[4]

The earthquake prediction issued by Dr. Whitcomb on April 15 was not his first. In November 1973 he correctly forecast that an earthquake would soon occur near Riverside, Calif.—it did on Jan. 30, 1974. There have been at least two other instances of U.S. scientists successfully predicting earthquakes, and numerous instances of detecting precursory signals from data they studied after earthquakes occurred. The most recent of the two predictions was made by geologists from the Geological Survey's Center for Earthquake Research in Menlo Park, Calif. They told a gathering of their colleagues on Nov. 27, 1974, that an earthquake would take place in nearby Hollister within days. Early the next morning the earthquake occurred.

The other forecast, the first scientifically valid one in this country, occurred not in seismically active California but in the

---

[4] Quoted in the *Los Angeles Times*, May 28, 1976.

Adirondack Mountains of New York. On Aug. 2, 1973, Dr. Yash P. Aggarwal, then a graduate student at Columbia University's Lamont-Doherty Geological Observatory, predicted that a minor tremor registering 2.5 to 3.0 on the Richter Scale would occur in two to four days near Blue Mountain Lake. As Aggarwal was sitting down to dinner two days later, the earth rumbled beneath his feet. "I could feel the waves passing by," he later recalled, "and I was jubilant."[5]

Concern Over Reaction to Official Forecasting

The question of when to make a prediction public has been vigorously debated in recent months. It has been argued that there should be sufficient delay to permit local authorities to prepare recommendations to the public. The respected British journal *Nature* came out recently for an extended delay. Predicting public reaction to an earthquake warning, it said, "seems to be every bit as difficult as predicting the earthquake itself."[6] Some scientists say that earthquake forecasts still are too unreliable to be made public. They are afraid that a false alarm might diminish public response to a later valid prediction.[7]

On the other hand, Dr. Ralph H. Turner, a sociologist at the University of California at Los Angeles, told the participants at the San Francisco conference on earthquake warning and response that "you simply have an absolutely unmanageable and uncontrollable situation if you start to prevent people who, for whatever reason, honestly believe that they have evidence, from releasing that information." It is bound to leak out, he said, embarrassing those who withheld it.

At the San Francisco conference, Geological Survey Director McKelvey unveiled a tentative federal plan for making earthquake predictions public. The plan calls for creation of an Earthquake Prediction Council, to be composed of five to ten Geological Survey scientists and outside specialists who would review predictions made by individual scientists and agencies. Their evaluation of the predictions would go to Geological Survey headquarters at Reston, Va., where a decision would be made whether to issue an "advisory notice" to the governor of the threatened state and to such agencies as the Federal Disaster Assistance Administration and the Defense Civil Preparedness Agency. The public either would be notified immediately or after a "short" delay. "It may be judged," McKelvey stated, "that the negative impact of a prediction could be lessened if responsible state and federal officials received prior notice."

---

[5] Quoted in *Time,* Sept. 1, 1975.

[6] Quoted by Walter Sullivan in "Forecasting Disaster," *The New York Times Magazine,* April 25, 1976, p. 54.

[7] For a discussion of public policy implications, see p. 137.

McKelvey interprets the Federal Disaster Relief Act of 1974 as placing responsibility on state and local officials for issuing any public warning. But according to *Science* magazine writer Deborah Shapley, the law's language is ambiguous and local officials, reluctant to alarm the public and local businesses, could argue that this still is a federal responsibility. Another criticism of the plan is that, officially at least, it covers only the work of Geological Survey scientists, although McKelvey voiced the hope that "other scientists would be willing to enter the system." So far most of the successful earthquake predictions have been made by university scientists. "The net result," Shapley wrote, "is that it is still unclear how—and whether—the public will be officially warned of a possible impending quake."[8]

Development of Earthquake Prediction Techniques

Scientists long have been making educated guesses about whether the probability of an earthquake is high or low in various places. California, for example, has not had a major earthquake since the San Francisco disaster of 1906, and for decades scientists have been saying that the state is long overdue for another one. Within the past five years, they have become convinced that the time, place and magnitude of earthquakes can be predicted within fairly close limits.

Methods used in the past to identify high risk areas depended largely on the incidence of quakes and the mapping of fault systems. Methods used now rely primarily on premonitory signs that occur in advance of a quake. Scientists working in widely separated parts of the world have identified several measurable physical, chemical and electrical changes that could indicate an imminent earthquake. Most of the signs are related to the increasing strains within the earth's crust that lead to a violent break.

It is unlikely that all of them occur before every earthquake, and they may differ from region to region. The earthquake prediction system that ultimately evolves "probably will not be based simply on one or two premonitory signs," said *New York Times* science editor Walter Sullivan. "Rather, the diagnosis of an approaching quake may be made in terms of many indicators, much as a physician diagnoses a disease on the basis of many tests."[9] Signs that have been identified with increased underground strain include the following:

Very slight changes in the tilt or elevation of the landscape near the threatened site.

Changes in the velocity at which sound waves traverse deep rock in the threatened area.

---

[8] Deborah Shapley, "Earthquakes: Los Angeles Prediction Suggests Faults in Federal Policy," *Science*, May 7, 1976, p. 537.

[9] Walter Sullivan, *op. cit.*, p. 50.

Changes in the electrical conductivity of crustal rock in the area.

Changes in the configuration of the earth's magnetic field.

Appearance of unusually large amounts of radon gas—a radioactive substance—in water from deep wells in the area.

Changes in the water level or temperature in the deep wells.

Increase in minor earthquake activity in the area.

Most of the precursory effects seem to be related to a phenomenon called dilatancy. Briefly stated, the theory is that accumulating pressure on subterranean rocks causes cracks to open—dilate—to relieve the stress. Water from surrounding rocks flows into the new fissures, saturating them. This restores the pressure and triggers the quake. Many seismologists believe the rise and fall of pressure follow a pattern that can be related to the time, size and location of the ensuing quake.[10]

China's Progress in Detecting Potential Tremors

Much of the research in earthquake forecasting is being conducted in the Soviet Union, in Japan, and particularly in China. China's long history of earthquake disasters provided strong motivation for perfecting an earthquake prediction capability. The impetus for the current research program came in March 1966 when two killer earthquakes struck Hopeh Province.

When a delegation of 10 American scientists toured Chinese earthquake centers in October 1974, they were astonished to learn that the country had some 10,000 scientists, engineers, technicians and other workers engaged in earthquake-prediction research. That was more than 10 times the number of such workers in the United States. An unusual aspect of the Chinese program was the use of amateurs, mostly students and peasants, whose responsibilities included collecting data pertinent to quake prediction, operating seismological instruments in remote areas, and educating the local people about earthquakes.

China was reported to be operating 17 fully equipped seismograph stations which, in turn, receive data from 250 auxiliary stations and some 5,000 observation points—some of which are simply wells where the radon content of the water is measured. The Chinese program, according to Dr. Frank Press, the U.S. delegation leader, encompasses every prediction method that has ever been suggested in any part of the world.[11] It

---

[10] See Christopher H. Scholz, Lynn R. Sykes, and Yash P. Aggarwal, "Earthquake Prediction: A Physical Basis," *Science,* Aug. 31, 1973, p. 809. The authors were geologists at the Lamont-Doherty Geological Observatory at Columbia University.

[11] See articles by Press: "Earthquake Prediction," *Scientific American,* May 1975, p. 21, and "Earthquake Research in China," *EOS,* November 1975, pp. 838-871. *EOS* is published by the American Geophysical Union.

includes such sophisticated techniques as the use of laser beams to measure distance and thus detect tiny deformations of the landscape.

The Chinese also pay close attention to oddities of animal behavior that seem to precede earthquakes. They told the visiting Americans that snakes, lizards and small mammals leave their underground burrows before an earthquake occurs and that insects congregate in huge swarms near seashores, cattle seek high ground, wild fowl leave their usual habitats and domestic animals become agitated. While many American scientists tend to view such reports with suspicion, Ruth B. Simon of the Geological Survey's National Earthquake Information Service in Denver writes that reports of unusual animal behavior prior to the occurrence of earthquakes are recorded in popular publications dating back as far as 1784.[12]

Plea for More Funds for American Program

Many American scientists involved in seismic research say that earthquake-prediction technology is hampered by insufficient funding in the United States. "It is simply a matter of two few methods being tested in too few places," Dr. Press has said.[13] According to his calculations, an additional $30-million a year is needed to make earthquake prediction a reality within a decade. The United States currently spends about $5-million a year on prediction research; legislation before Congress would—if passed—increase the amount but still fall short of $30-million a year.[14]

Sen. Alan Cranston (D Calif.), a chief sponsor of the authorization bill, told the Senate: "The United States today faces the greatest potential danger from earthquakes that we have ever faced before. It is only in the last decade or so that our population has become concentrated in major cities and along our coastal regions, and major construction has occurred on landfill and other unstable soils." If California were to experience today an earthquake comparable to the 1906 San Francisco quake, Cranston said, deaths could number in the tens of thousands and the property damage could exceed $20-billion. Some 70 million Americans live with a significant risk to their lives and property from earthquakes, according to the National Academy of Sciences. Some 115 million others are exposed to a less significant, but not negligible, seismic risk. Only 8 per cent of Americans can safely ignore the earthquake hazard.[15]

---

[12] Department of the Interior News Release, Jan. 22, 1976.

[13] Frank Press, "Earthquake Prediction," *Scientific American*, May 1975, p. 18.

[14] The Senate voted on May 24, 1976, to authorize $75-million during the next three years for Geological Survey research, including earthquake forecasting. The House held hearings on the authorization bill the following month.

[15] Panel on the Public Policy Implications of Earthquake Prediction, "Earthquake Prediction and Public Policy," National Academy of Sciences, 1975, p. 20.

Earthquake Causes and Consequences

COMPARED TO OTHER seismically vulnerable regions, the United States has been extraordinarily lucky. In 200 years only 1,600 Americans have died in earthquakes. Throughout history some 74 million persons are known to have died as a result of earthquakes or the floods, fires and landslides they triggered. The earliest accurate recording of an earthquake and its toll was in the year 856 in Corinth, Greece. Some 45,000 persons were killed in that quake. China holds the record for the greatest toll from a single earthquake—830,000 Chinese were killed in the Shensi quake of 1556.

The first great earthquake to be observed in anything approaching a scientific manner took place in Lisbon, Portugal, on Nov. 1, 1755. There were three principal shocks, the first and strongest of which lasted six or seven minutes, an unusually long duration. It was All Saints Day, and thousands were crushed as churches and cathedrals collapsed. About 60,000 of the city's 235,000 inhabitants died and all large public buildings and 12,-000 dwellings were demolished. But Portugal was not the only country to feel the quake's impact.

> As far away as Scotland, the waters of Loch Lomond were agitated; canals in Amsterdam and Rotterdam shook until many large ships snapped their cables. In Morocco, 10,000 people were killed.... In Bohemia, 1,380 miles from Lisbon, medicinal springs became turgid and muddy and then ceased to flow. The seismic sea waves swept over the North African coasts; fjords and lakes in Norway and Sweden were disturbed. Scores of fish were left high and dry on the streets of Madeira. The land west of Lisbon, near Cadiz, was permanently elevated by several feet.[16]

The first technically documented earthquake occurred in Assam, India, on June 12, 1897. The credit for the scientific investigation goes to R. D. Oldham, who was then director of the Geological Survey of India. He carried out a thorough investigation of the earthquake's effects and a monograph he wrote is still a valuable sourcebook in seismology. The quake, which had a magnitude of 8.7, killed about 1,500 people and demolished Shillong, a city that has been rebuilt and is today a center of seismological studies in India.

In this century, the toll has been immense. In 1908 a quake took 50,000 lives in Messina, Italy. Some 180,000 were killed in Kansu Province, China, in 1920 and 70,000 in a second earthquake in 1932. In Japan, where seismographs record a quake

---

[16] Eloise Engle, *Earthquake!* (1966), p. 31.

Earthquakes 1976

Earthquakes have been front page news frequently this year, beginning with the tragic Guatemalan quake on Feb. 4 which claimed some 22,000 lives. May was a particularly active month. Five tremors of serious magnitude were reported, including the May 6 quake in northern Italy which killed almost 1,000 persons, injured 1,700 and left 100,000 homeless, and the May 16 shock in Soviet Central Asia which took an undetermined number of lives.

About 9,000 people are believed to have been killed on June 26 by an earthquake and subsequent landslides in Indonesia's remote West Irian Province on the island of New Guinea. On July 11 two major earthquakes shook the sparsely populated jungles of the Panama-Colombia border region. There were no immediate reports of injuries.

Despite the apparent increase in world-wide earthquake activity, the U.S. Geological Survey reports that the number of major earthquakes in 1976 actually is running behind the long-term yearly average. So far this year the world has experienced 21 significant earthquakes. But only 10 of the 21 have been classified as "major" quakes—those registering 7.0 or higher on the Richter Scale. Normally 16 to 18 major earthquakes are expected each year, although in each of the past three years only about a dozen occurred.

every hour, 143,000 people were killed on Sept. 1, 1923, by a giant quake that devastated the cities of Tokyo and Yokohama. The Chilean earthquakes of 1960 brought death to 2,000 people, injury to 5,000 more and destruction to 50,000 homes.[17] The Peruvian earthquake of May 31, 1970, the worst in South American history, left 70,000 dead and 800,000 homeless. More recently, an earthquake in Managua, Nicaragua, on Dec. 23, 1972, resulted in 10,000 deaths, 5,000 injuries and property loss equivalent to the annual gross national product; the earthquake in Guatemala this year *(see box above)* caused 22,000 deaths.

Memorable Earthquakes in the United States

Probably the most destructive earthquakes in U.S. history were those that occurred in the vicinity of New Madrid, Mo., in the winter of 1811-1812. Residents of that small Mississippi River community were awakened on the morning of Dec. 16, 1811, when the trembling earth shook them from their beds. There followed a series of earthquakes which continued, several each day, until February 1812. Shocks were felt as far away as Boston, New Orleans and Detroit. The significance of these quakes was not in the number of lives lost, which were few in that sparsely populated area, but in what they did to the land.

---

[17] In addition to the series of intense earthquakes, Chile suffered tidal waves of up to 24 feet, volcanic eruptions and landslides. The tidal wave raced across the Pacific, causing 61 deaths in the Hawaiian Islands and 180 in Japan.

Most of the spectacular effects were confined to the Mississippi River and its tributaries and to the bottom lands. These bottom lands consist of a thin layer of alluviam overlying a water-saturated bed of sand. This unconsolidated material was radically modified by the earthquake, sinking in many places to form swamps and in other places to form lakes, permanently water-covered, where forests had grown before.[18]

While the New Madrid quakes may have been the most extensive in American history, the San Francisco earthquake of April 18, 1906, remains the most famous. Fires caused by overturned stoves, falling chimneys and ruptured gas pipes destroyed much of the city. But in relation to its population, the hardest-hit town was Santa Rosa, north of San Francisco. Santa Rosa's death toll was more than 75; if the loss of life in San Francisco had been in proportion, it would have been 8,000 instead of 450.

The Alaskan earthquake of March 27, 1964, caused damage throughout an area of 50,000 square miles and tremors were felt in an area of one million square miles. A report issued on Nov. 11, 1967, by the Environmental Science Services Administration of the U.S. Department of Commerce listed some of the more noteworthy effects of the Alaska quake: 131 persons died in Alaska and along the Pacific Coast; a seismic wave was recorded in Antarctica 22½ hours after the earthquake; shock waves oscillated water as far away as Key West, Fla.[19]

The last big earthquake in the United States occurred Feb. 9, 1971, when Southern California was jolted by a tremor centered in the San Fernando Valley near Los Angeles. The quake, which registered 6.6 on the Richter Scale, killed 64 persons, including 44 patients of the Veterans Administration Hospital at Sylmar. Damage at the Veterans Hospital presented a classic picture of building techniques. The center structures, built in 1926, collapsed "like smashed orange crates," according to an observer. The outer structures, built in 1937 and 1947 after earthquake-resistant designs were included in building codes, escaped without any significant structural damage.

Some Early Attempts at Earthquake Forecasting

Earthquake prediction was a constant preoccupation for early soothsayers, astrologers and prophets, and there are many instances recorded in history of destructive earthquakes having been forecast. The earthquake of 1042 in the Persian village of Tabriz was predicted by the chief astrologer, who tried in vain to persuade the people to leave. Earthquakes were a frequent happening in Tabriz, yet, according to Professor Nicholas N.

---

[18] John H. Hodgson, *Earthquakes and Earth Structure* (1964), p. 8.

[19] The quake registered as high as 8.6 on the Richter Scale.

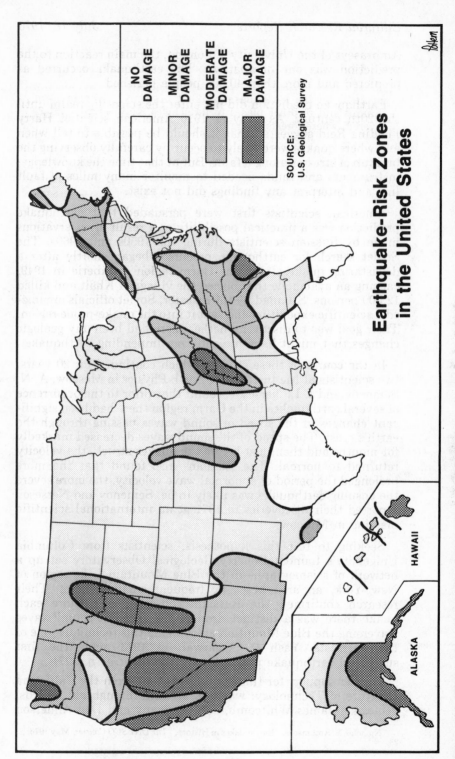

Earthquake-Risk Zones
in the United States

	NO DAMAGE
	MINOR DAMAGE
	MODERATE DAMAGE
	MAJOR DAMAGE

SOURCE:
U.S. Geological Survey

ALASKA

HAWAII

Ambraseys of the University of London, the main reaction to the prediction was one of apathy.[20] The earthquake occurred as predicted and more than 40,000 people perished.

Earthquake prediction did not enter the scientific realm until the 20th century. As early as 1910 American scientist Harry Fielding Reid suggested that it should be possible to tell when and where quakes were likely to occur by carefully observing the buildup of stresses along a fault. But at that time the knowledge, instruments and funds needed to monitor many miles of fault line and interpret any findings did not exist.

American scientists first were persuaded that earthquake prediction was a practical possibility as a result of observations made by Russian scientists during the 1950s and 1960s. The Soviet search for earthquake precursor began shortly after a devastating quake struck the Garm region of Siberia in 1949, causing an avalanche that buried the village of Khait and killed 12,000 persons. Stunned by the disaster, Soviet officials organized a scientific expedition and sent it into the quake-prone region. Their goal was to discover whether there had been any geologic changes that might have indicated an impending earthquake.

In the course of these studies, which continued for 20 years, two scientists at the Institute of Earth Physics in Moscow, A. N. Semenov and I. L. Nersesov, found that prior to the occurrence of several earthquakes in the Garm region there had been significant changes in the speed of sound waves passing through the earth's crust. The speed of the sound waves decreased markedly for months and then, just before the quakes struck, the velocity returned to normal. The Russians also found that the more prolonged the period of abnormal wave velocity, the more severe the ensuing earthquake was likely to be. Semenov and Nersesov reported their discoveries in 1971 at an international scientific meeting in Moscow.

Seeking to test this hypothesis, scientists from Columbia University's Lamont-Doherty Geological Observatory set up a network of seismographs in the Blue Mountain Lake region of New York, an area beset by frequent small tremors. Their research confirmed the Russians' observations: before each quake there was a distinct drop in the speed of sound waves traversing the Blue Mountain area. Using this technique, one of the scientists, Yash P. Aggarwal, in 1973 made the first successful earthquake prediction in America *(see p. 127)*.

Further support for this hypothesis came from the California Institute of Technology, where, searching through old records, geologists James Whitcomb, Jan Garmany and Don Anderson

---

[20] Nicholas N. Ambraseys, "Earthquakes in History," *The UNESCO Courier*, May 1976, p. 28.

found evidence of a distinct drop in sound wave speeds predating the 1971 San Fernando Valley earthquake by three and a half years. The waves had returned to their normal velocity a few months before the tremor. In August 1973, Japanese scientists told a scientific meeting in Boulder, Colo., that they had detected the same phenomenon in records made before four Japanese earthquakes that occurred in the 1960s.

Meanwhile, other early warning signals had been detected—increased emissions of radon gas, changes in electrical conductivity of deep rock, changes in the tilt or elevation of the landscape. Searching for a link between the various precursors, American scientists returned to a theory first postulated by geologist William F. Brace of the Massachusetts Institute of Technology in the mid-1960s. Brace had discovered that the physical properties of rocks change drastically when they are placed under great stress. Their resistance to electricity increases and seismic waves passing through them slow down.

These changes seemed related to a phenomenon known as dilatancy—the property some materials have of expanding in volume when their shape or structure is altered under stress. Brace even suggested at the time that the physical changes associated with dilatancy might provide warning of an impending earthquake, but neither he nor anyone else was quite sure how to proceed with his proposal. "Dilatancy was, in effect, put on the shelf."[21] The Russian discoveries reawakened interest in the subject. Among the first scientists to explain the existence of earthquake precursors by the theory of dilatancy were Amos M. Nur of Stanford University in 1972, and Christopher H. Scholz, Lynn R. Sykes and Yash Aggarwal of Columbia University in 1973. The theory of dilatancy now is supported by most American earthquake specialists.

Uncertainty About Public Reaction

EARTHQUAKE FORECASTING, like any new technological capability, will have both positive and negative effects. Along with the obvious benefits will come a host of side issues and new problems. "The situation is in some ways similar to that in 1939 when nuclear fission suddenly became a reality," the editor of the British journal *Nature* wrote in 1973. "The prospects for society are neither uniformly good nor uniformly bad and there is still time, but relatively little, to explore ways in which good prospects can be encouraged and bad ones minimized."[22]

[21] "Forecast: Earthquake," *Time*, Sept. 1, 1975, p. 38.
[22] Quoted by Kendrick Frazier in *Science News*, Oct. 20, 1973, p. 243.

Many public officials fear that earthquake predictions and warnings will lead to a mass migration of people and industries, a drastic drop in real estate values, reduced tax revenues, and widespread unemployment for remaining residents. Among the first to say that earthquake predictions could do more harm than good was Dr. Garrett Hardin, professor of human ecology at the University of California at Santa Barbara. In an essay written nearly 10 years ago, Professor Hardin argued that no further research be done on the subject. His primary concern was the psychological effects of earthquake forecasting.

> We who live in earthquake zones feel little anxiety about the future earthquakes that will come at completely unforeseen times; Freudian denial arms us against their psychological threat. The development of an ability to predict a quake months away with considerable reliability would tend to remove seismic phenomena from the category of fate. Being able to predict an earthquake, we would feel that we should do something about it; but our relative impotence would generate anxiety.[23]

Such concerns prompted the National Academy of Sciences in 1974 to name a panel to study the public policy implications of earthquake prediction. In its final report, released Aug. 25, 1975, the committee conceded that "under the worst combination of inaccurate prediction and an inappropriate public response, the prediction and the quake together might even be more costly than an unpredicted quake would have been." Nevertheless, the committee strongly recommended continued research into earthquake prediction. Forecasting would clearly save lives, the panel stated, and that is the "highest priority."[24] According to the report, Americans are unlikely to panic at an earthquake forecast. They are more likely to adopt a business-as-usual attitude.

Public Response to the Recent California Forecast

Social scientists were anxious to observe the economic effect of the recent earthquake prediction for Los Angeles. So far the fear of falling land values and public panic have not materialized. Except for a sharp rise in the sale of earthquake insurance, the forecast has had little socio-economic impact, according to Dr. Eugene Haas, a sociologist with the Institute of Behavioral Sciences at the University of Colorado. Dr. Haas, who has been studying the social aspects of natural hazards for more than a decade, found no signs of any change in property values, buying or selling of homes, or changes in lending policies for building or buying homes.[25]

---

[23] Essay reprinted in Garret Hardin, *Stalking the Wild Taboo* (1973), pp. 123-134.
[24] Panel on the Public Policy Implications of Earthquake Prediction, *op. cit.*, p. 3.
[25] Interviewed in *The New York Times*, May 15, 1976.

Plate Tectonics Theory

Within the past 15 years, scientists have developed a new theory called plate tectonics which offers a comprehensive explanation for the underlying cause of earthquakes. The theory of plate tectonics is an updated version of the theory of continental drift first proposed by the German meteorologist Alfred Wegener in the second decade of this century.

According to the theory of plate tectonics, the earth's crust is not a single, uniform layer, but a discontinuous series of about a dozen huge, shifting plates, each 30 to 90 miles thick. Floating on the earth's semimolten mantle and propelled by as yet undetermined forces, the plates are in constant motion. As a result of friction between plate movements, stress builds up in the earth and rocks. Eventually the rock fractures, and energy is released. It is the sudden release of this pent-up energy that causes earthquakes.

Soon after Dr. Whitcomb made his earthquake prediction public, he was denounced by a member of the Los Angeles City Council, who threatened to sue him for "harm to San Fernando Valley property values." The *Los Angeles Times* reported, however, that "most Southern Californians are treating the latest earthquake prediction the same way they seem to have treated all the rest—with a mixture of skepticism, caution and prayer."[26]

Recommendations From National Academy of Sciences

The National Academy's panel concluded that forecasting will benefit the public only if appropriate social, economic, engineering and legal actions are taken before an earthquake occurs. The committee expressed concern that the costs involved might impel public officials to do nothing. The committee recommended that earthquake predictions be made public promptly and without respect to policy considerations. To guard against false predictions, the group said a panel of impartial experts should be set up to review all predictions and gauge their scientific validity. The committee warned that the responsibilities of federal, state and local officials for issuing earthquake warnings are "dangerously" undefined and should be clarified immediately. Once a prediction has been issued, the panel said, the community should prepare emergency services, identify escape routes and communication lines, and assure adequate and appropriately dispersed emergency equipment and supplies.

Most casualties from earthquakes occur when buildings collapse, fires start or dams burst. Therefore the committee recommended that a community use the time provided by the prediction to reinforce or demolish vulnerable buildings. As the

---

[26] *Los Angeles Times*, April 22, 1976.

time of a predicted quake draws near, the committee suggested removing people from buildings likely to collapse, lowering the water level in dams, shutting off gas lines that could burst and cause fires, and closing down nuclear power plants.

Although the committee concluded that evacuation of entire populations is impractical, it said that in some cases Americans might follow the example of the Chinese, who often leave their buildings and live in tents in the countryside when an earthquake is predicted. The panel also said that the success of the hazard-reduction program will be greatest in communities where standards for earthquake-resistant construction are already strictly enforced, where some continuing identification of safe and unsafe structures is maintained, and where land-use management has systematically taken seismic risk into account.

The panel said that all phases of earthquake prediction and response will be plagued initially by legal uncertainties. It recommended that a legal inquiry be made to determine if, under the Disaster Relief Act of 1974, low-interest loans and relief money usually made available to disaster-hit communities could be released before an earthquake occurred. This would enable the endangered community to tear down unsafe buildings, shore up marginal structures, strengthen or evacuate hospitals and nursing homes, and upgrade emergency aid and medical services. If this is not possible under existing law, the committee said, a new law should be passed.

Possibility of Earthquake Control or Modification

Buoyed by their progress in earthquake forecasting, scientists are looking ahead toward another goal: earthquake control—the "ultimate step in eliminating earthquake disasters."[27] The possibility of controlling or modifying earthquakes arose in 1966 from a chance discovery. Geological Survey scientists in the Denver area observed that the forced pumping of lethal wastes from the manufacture of nerve gases into deep wells at the Army's Rocky Mountain Arsenal coincided with the occurrence of hundreds of small quakes near Denver. After the Army suspended the waste-disposal program, the number of quakes declined sharply. The earthquakes apparently had been triggered by the injection of fluid into stressed rock.[28] If quakes could be "turned on" in this manner, the scientists wondered, could they be "turned off" by removing water?

The Geological Survey began testing this theory in 1969 at the Rangely Oil Field in northwest Colorado. There the Chevron

---

[27] So described by Dr. Carl Kisslinger, professor of geological sciences and director of the Cooperative Institute of Research in Environmental Sciences at the University of Colorado, in testimony on Feb. 19, 1976, before the Senate Subcommittee on Oceans and Atmosphere.
[28] See "Chemical-Biological Weaponry," *E.R.R.*, 1969 Vol. I, p. 454.

Oil Co. had set off small quakes when it injected water under pressure into oil wells, a technique used to recover more oil. With Chevron's permission, the Geological Survey pumped the water out of certain wells to see what would happen. When that was done earthquake activity in the area virtually ceased.[29]

Two geologists of the federal agency, C. B. Raleigh and James Dietrich, subsequently suggested that the pressure in rocks near active faults could be modified, causing the release of the stored energy in a relatively safe and uncontrolled way. To test this theory, they propose that five wells be drilled parallel to an active fault in an area where there are indications of strain buildup, such as along the San Andreas Fault. Fluids, primarily water, would be pumped out of all but the center well. This would have the effect of drying the rock in the vicinity of the outer wells and making it bind together more tightly, thus strengthening its resistance to movement (slippage) which could result in an earthquake.

Water would then be pumped down the center well. The lubricating effect of this injection of fluid would cause the rock near the center well to slip, thus producing earthquake activity. In theory, the earthquake shock would be confined to the relatively small area surrounding the center well by the neighboring zones of strengthened rock. If the process proved to be successful, it then could be extended to other areas along the fault. The minor quakes that would be produced would release the stored energy in the vicinity of the fault in a controlled way, thus minimizing the chances of a large natural earthquake.[30]

The scientific community generally supports this idea but warns that such experiments must be undertaken with caution lest they trigger a major earthquake. Some scientists believe that a technical capability in earthquake modification could be developed before earthquake prediction techniques are perfected. One problem is money. Geologists Raleigh and Dietrich estimate that the well-drilling experiment would cost $1-billion to $2-billion. Today the prospect of such lavish financing is remote. As Frank Press puts it, "How does one sell preventive medicine for a future affliction to government agencies beleaguered with current illnesses?"[31] Ironically, the one event that would release abundant resources for a large-scale earthquake-control program is the very disaster scientists are trying to prevent—a major earthquake striking a highly populated area without warning.

---

[29] See C. B. Raleigh, J. H. Healey and J. D. Bredehoef, "An Experiment in Earthquake Control at Rangely, Colorado," *Science*, March 26, 1976, pp. 1230-1237, and Walter Sullivan, "Turning Off an Earthquake." *The New York Times Magazine*, April 25, 1976, p. 56.

[30] See Russell Robinson, "The Rumbles on Seismos," *Natural History*, January 1972, p. 44.

[31] Frank Press, *op. cit.*, p. 23.

Selected Bibliography

Books

Engle, Eloise, *Earthquake!,* The John Day Co., 1966.
Halacy, D. S. Jr., *Earthquakes: A Natural History,* Bobbs-Merrill, 1974.
Iacopi, Robert, *Earthquake Country,* Lane Books, 1971.
Smith, Peter J., *Topics in Geophysics,* MIT Press, 1973.
White, Gilbert F., and J. Eugene Haas, *Assessment of Research on Natural Hazards,* MIT Press, 1975.

Articles

Anderson, Alan Jr., "Earthquake Prediction," *Saturday Review,* February 1973.
Canby, Thomas Y., "Can We Predict Quakes?" *National Geographic,* June 1976.
Hammond, Allen L., "Earthquakes: An Evacuation in China, A Warning in California," *Science,* May 7, 1976.
Koughan, Martin, "Goodbye, San Francisco," *Harper's,* September 1975.
Press, Frank, "Earthquake Prediction," *Scientific American,* May 1975.
Purrett, Louise, "The Possibilities of Earthquake Prediction," *Science News,* Feb. 20, 1971.
Raleigh, C. B., et al., "An Experiment in Earthquake Control at Rangely, Colorado," *Science,* March 26, 1976.
Robinson, Russell, "The Rumbles on Seismos," *Natural History,* January 1972.
Scholz, Christopher, et al., "Earthquake Prediction: A Physical Basis," *Science,* Aug. 31, 1973.
Shapley, Deborah, "Earthquakes: Los Angeles Prediction Suggests Faults in Federal Policy," *Science,* May 7, 1976.
Sullivan Walter, "Forecasting Disaster," *The New York Times Magazine,* April 25, 1976.
The UNESCO Courier, May 1976 issue.
U.S. Department of the Interior Geological Survey, *Earthquake Information Bulletin,* selected issues.
Young, Patrick, "Seismic Sleuthing," *Saturday Review/World,* Feb. 23, 1974.
"Forecast Earthquake," *Time,* Sept. 1, 1975.

Studies and Reports

Editorial Research Reports, "Earthquakes: Causes and Consequences," 1969 Vol. I, p. 203.
Panel on the Public Policy Implications of Earthquake Prediction, "Earthquake Prediction and Public Policy," National Academy of Sciences, 1975.
Subcommittee on Oceans and Atmosphere of the Committee on Commerce, "Hearings on Earthquake Disaster Mitigation Act of 1975," Feb. 19, 1976.
U.S. Department of the Interior, "Earthquake Prediction—Opportunity to Avert Disaster," Geological Survey Circular 729, 1976.
—— "Goals, Strategy and Tasks of the Earthquake Hazards Reduction Program," Geological Survey Circular 701, 1974.

SOLAR ENERGY

by

John Hamer

Editor's Note: Since this report was published, the Carter administration and Congress have taken several steps to foster applications of solar energy. The Carter energy plan, announced in April 1977, included provisions for tax credits for the installation of solar energy equipment. The energy plan that Congress passed in October 1978 provided a credit of 30 percent of the first $2,000 spent and 20 percent of the next $8,000—with a maximum credit of $2,200—for homeowners who install solar, wind or geothermal energy equipment.

The same month Congress passed a bill setting up a $1.5 billion, 10-year federal program to promote development of photovoltaic cells that convert sunlight into electricity. The program is intended to double annual production of photovoltaic cells and reduce the cost tenfold.

Finally, William A. Shurcliff, mentioned on p. 148, has stopped making his annual survey of solar buildings in the United States. There are, he said, just too many solar buildings going up to keep track of them.

SOLAR ENERGY

SOLAR ENERGY is by far the most abundant energy source available to Earth. The sun generates such an enormous amount of energy that the facts and figures are almost incomprehensible. Only an infinitesimal fraction of the sun's radiant energy strikes this planet, but our share still equals about 180 trillion kilowatts of electricity, or more than 25,000 times the world's present industrial power capacity. A few comparisons help to illustrate the awesome potential of solar energy. The energy in the sunlight falling on the surface of Lake Erie in a single day is greater than current annual U.S. energy consumption. The amount of solar radiation striking only 1 per cent of the nation's land area each year is more than projected national energy needs to the year 2000. The solar energy reaching the surface of the entire United States annually is greater than the total amount of fossil-fuel energy that, scientists say, will *ever* be extracted in this country.

The sun is already the indirect source of most of the energy used on earth—from wood, wind and falling water to the coal, oil and natural gas deposits formed centuries ago. But today, with fossil-fuel resources rapidly being depleted, nuclear energy plagued by cost and safety problems, and hydropower and geothermal resources limited, interest in wider use of solar energy is soaring. Solar power is seen as the clean, safe, pollution-free and virtually inexhaustible energy source that can meet the nation's—and the world's—energy supply needs for the foreseeable future. A wide variety of technologies are being pursued, from direct use of solar energy to heat water and buildings or generate electricity to indirect use through the wind, ocean thermal layers or bioconversion of organic material *(see boxes, pp. 148 and 150)*.

But there are many problems. For one thing, sunlight is so diffuse that collecting it and concentrating it present serious difficulties. Solar energy varies greatly with latitude, season, time of day and weather conditions. Moreover, it cannot be converted to useful power at 100 per cent efficiency, and it cannot be stored easily for later use or transported to other areas. There are numerous technical problems with most existing solar-energy equipment, although progress has been rapid in recent years and more refinements or scientific breakthroughs seem imminent. Perhaps the greatest barriers to the acceptance of solar energy are political, social and economic. But none of these problems appears insoluble and it is increasingly likely that solar energy will emerge in the decades ahead as a major energy source for the United States and for other nations.

The unanswered questions are: How much will this energy cost and how soon will it be available? So far solar energy has not been economically competitive with cheap fossil fuels. And the future cost of solar energy is even more unpredictable than the price of oil, which has risen far beyond most expectations in recent years. As for availability, estimates vary. Proponents of solar energy say it could meet 50 to 100 per cent of the nation's energy needs by the end of the century, while skeptics contend it will provide only 5 to 10 per cent. The Energy Research and Development Administration, which has primary responsibility for federal solar-energy programs, is in the middle, predicting that solar energy will provide about 25 per cent of the nation's needs by the year 2020.

Existing Uses for Heating Water and Buildings

The first widespread use of solar energy in this country probably will be for heating water and for heating—and possibly cooling—buildings. Solar water heaters have been around for decades and many are still in use today. They are fairly simple devices, and most units work about the same way. A sheet of metal—often copper—and a series of metal tubes are painted black to absorb the sunlight, then covered by glass to trap the heat like a greenhouse. The unit is mounted on the roof, facing the sun. Water is pumped through the tubes where it collects heat, then is stored in an insulated tank much like conventional hot-water heaters. The water temperature commonly reaches 100 to 200 degrees on sunny days. The devices also can be used to heat water for swimming pools.

For space heating, the principle is similar. The sun's heat is transferred either to water or to air, which is then pumped or blown into a heating system's radiators or ducts. The heat can be stored in hot-water tanks or in a bed of hot rocks for release at night or in bad weather. At the beginning of this decade only a handful of homes and buildings in the United States had solar-heating units, but today there are more than 200 and new systems are being installed all the time. Many more are in the planning stages. Solar energy systems now are operating in private homes, government buildings, commercial establishments and public schools. Technically at least, solar energy also can be used for air-conditioning systems that, much like a refrigerator, pump a coolant. But more research is needed to improve their efficiency.

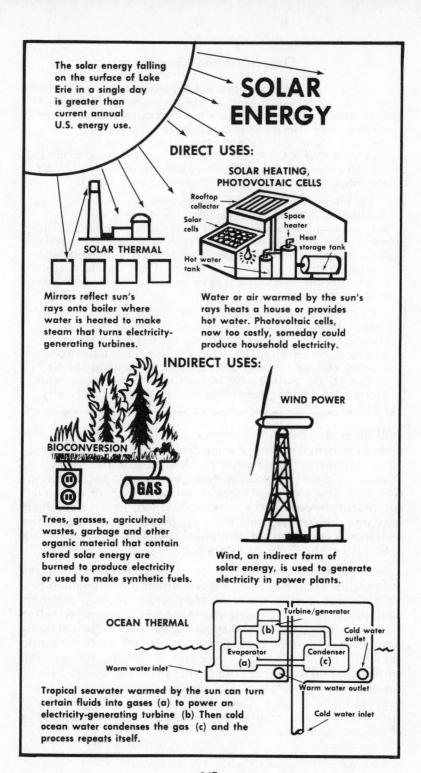

The solar energy falling on the surface of Lake Erie in a single day is greater than current annual U.S. energy use.

SOLAR ENERGY

DIRECT USES:

SOLAR HEATING, PHOTOVOLTAIC CELLS

Rooftop collector

Solar cells

Space heater

Heat storage tank

Hot water tank

SOLAR THERMAL

Mirrors reflect sun's rays onto boiler where water is heated to make steam that turns electricity-generating turbines.

Water or air warmed by the sun's rays heats a house or provides hot water. Photovoltaic cells, now too costly, someday could produce household electricity.

INDIRECT USES:

BIOCONVERSION

GAS

WIND POWER

Trees, grasses, agricultural wastes, garbage and other organic material that contain stored solar energy are burned to produce electricity or used to make synthetic fuels.

Wind, an indirect form of solar energy, is used to generate electricity in power plants.

OCEAN THERMAL

Turbine/generator

Cold water outlet

Evaporator (a)

Condenser (c)

(b)

Warm water inlet

Warm water outlet

Cold water inlet

Tropical seawater warmed by the sun can turn certain fluids into gases (a) to power an electricity-generating turbine (b) Then cold ocean water condenses the gas (c) and the process repeats itself.

Ocean Thermal Gradients

Temperature differences between sea water at the surface and at great depths—ocean thermal gradients—are another indirect form of solar energy. This source has been called potentially the greatest of all. About half of the radiant solar energy that strikes the earth annually lands in the tropics, where most of it acts to heat the surface of the oceans. This energy can be tapped by using the sun's stored heat together with the stored cold in water from the ocean depths *(see p. 147)*.

The process is fairly simple: An electricity-generating turbine is installed offshore in a facility capable of drawing warm seawater from the surface or colder water up from the depths. The warm water is used to warm a fluid such as propane or ammonia and turn it into a gas. The gas expands inside a container and provides the power to operate the turbine. Then, upon being cooled by the colder water, the gas again becomes a fluid to be reused over and over again.

The idea first was proposed in 1881 by the French physicist Arsene d'Arsonval, and an experimental model was built by Georges Claude of France in 1929, but the technology has received little attention since. The National Science Foundation has estimated that by the year 2000 thermal gradients in the Gulf Stream off the eastern coast of the United States alone could generate 5 per cent or more of the energy needed in this country.

William A. Shurcliff, a Harvard University physicist who makes an annual survey of solar-heated buildings in the United States, wrote last February: "A solar-heating industry is developing rapidly. Hardly a week goes by without the formation of a new company eager to sell collectors or associated equipment. Several giant corporations, too, have tossed their hats into the ring. Already there are over 40 companies in the United States offering water-type collectors for the solar heating of buildings, and there are a dozen offering air-type collectors."[1]

None of these ventures can be considered a great success today, although the future prospects are bright. Manufacturers of rooftop collector systems, each using slightly different equipment, are trying constantly to modify it to achieve better performance, longer durability and lower cost. "The competition involves, mainly, petty details," Shurcliff wrote. "No new principles of physics are expected, or sought. Rather, the emphasis is on salvaging a bit of otherwise lost energy here or there, reducing corrosion of a certain aluminum component, preventing a plastic sheet from warping or discoloring, eliminating a valve, finding cheaper materials and cheaper ways of fabricating them."

---

[1] William A. Shurcliff, "Active-Type Solar Heating Systems for Houses: A Technology in Ferment," *Bulletin of the Atomic Scientists,* February 1976, p. 30.

The solar-heating industry today has been compared to the automobile industry at the beginning of the century, waiting for a breakthrough that will bring rapid sales growth. However, it was mass production that enabled auto manufacturing to expand so rapidly, and most of the components of solar systems already are being mass produced. What will really help the industry are continued price increases for fuel oil, natural gas and electricity. The initial installation cost of solar units still is so high that it takes at least five and often 10 to 15 years before the cost is offset by savings on utility bills. As conventional utility costs increase, solar equipment will become more attractive economically.

There is hope for savings and efficiency in the combination of solar units with conventional systems. The heat pump, an electric-powered device already used in about one million American homes for heating and cooling, can be combined with solar rooftop collectors to lower heating bills significantly. The largest solar-heated building in the world, at New Mexico State University in Las Cruces, has such a system.

Converting Sunlight to Electricity; Cost Factors

Solar researchers also are giving considerable attention to the generation of electricity, especially since electricity is expected to provide more than half of this country's energy needs by 1985, compared to 25 per cent today. The two principal methods of converting sunlight to electricity are (1) through the use of photovoltaics, or solar cells, and (2) solar thermal conversion, the use of focused sunlight to heat water and drive steam turbines in an electric-power plant. The technology is not sufficiently advanced to make either method economically competitive with conventional ways of producing electricity. But the potential is great.

Photovoltaic cells have been used in America's space program for many years to provide energy in satellites, but the cost is extremely high. The Skylab space station ran entirely on solar cells, but at a cost of about $300,000 a kilowatt. Less intricate systems can be operated on earth for about $20,000 a kilowatt, still too high for all uses except in isolated places such as offshore oil rigs or remote communications stations. Still, the cost of solar cells has fallen to under $20 a watt from $200 a watt only five years ago, and further decreases are expected, since the technology is somewhat similar to the solid-state technology that has greatly reduced the cost of electronic calculators.

Solar cells have no moving parts. They consist mainly of two thin layers of material, one of them a semiconductor such as silicon and the other a metal such as aluminum or silver. A semiconductor can be treated so that when light strikes it

149

Energy From Bioconversion

A possible use of solar energy is the transformation of trees, plants and other organic material into useful power, a process known as bioconversion. Wood and dry animal manure have been burned for centuries as a small-scale source of heat, but today researchers are suggesting large-scale projects to produce electricity or gases for commercial or industrial use. The present annual production of "biomass" around the world is about 100 billion dry tons, which has an energy equivalent about six times greater than current energy use worldwide.

Converting this material into useful power is theoretically feasible, although the technology remains to be demonstrated practically. The National Science Foundation has recommended that pilot plants be built and tested using several bioconversion techniques, including the direct combustion of trees, certain grasses and other plants to make steam in electric power plants, and the conversion of agricultural wastes, municipal refuse, human sewage and animal wastes into methane gas or methyl alcohol.

Union Electric Co. in St. Louis already is using shredded garbage mixed with coal in power plants, and wastes in Baltimore and San Diego are being subjected to pyrolisis (chemical change induced by heat) to yield gases, oil and char for use as boiler, home-heating and motor fuels.

electrons flow across the two layers—the so-called photovoltaic effect—and generate current. This current is drawn off in wires to operate electric motors, light bulbs or other devices. Today photovoltaic flashlights, radios and television sets are available, but their price is much higher than their conventional counterparts. A panel of solar cells measuring about 10 feet by 40 feet could provide all the electricity needed by the average home, but it would cost more than $100,000.

Solar cells are expensive to manufacture because they are made from chemically pure silicon crystals that are sliced into ultrathin wafers by a precision, labor-intensive process. However, several companies are experimenting with methods of making long silicon crystal rods or ribbons that could be sliced in a mass-production technique. Solarex Corp. of Rockville, Md., and Mobil Tyco Laboratories of Waltham, Mass., are investing heavily in solar cells in the hope of making a research breakthrough. "The day may arrive when solar cells are delivered to a house like rolls of roofing paper, tacked on, and plugged into the wiring, making the home its own power station," John L. Wilhelm wrote in *National Geographic*.[2]

---

[2] John L. Wilhelm, "Solar Energy, the Ultimate Powerhouse," *National Geographic*, March 1976.

Another ambitious plan for using solar cells is put forward by Peter Glaser of Arthur D. Little Inc., an industrial consulting firm in Cambridge, Mass. Glaser suggests that a huge satellite in orbit could use solar panels several miles long to generate electricity and beam it back to earth in the form of microwaves to a receiving station. The microwaves would be converted into alternating-current electricity and distributed through utility grids. The plan, though futuristic, is technically feasible and is under study by such companies as Boeing, Grumman, Textron and Raytheon. Assuming the development of new space vehicles and methods for construction in orbit, Glaser believes the system could become operational by the year 2000.

Solar thermal conversion is still in a developmental stage in the United States, although small systems began operation in France and Mexico this year. They focus the sun's heat on a boiler to produce steam to run turbines that power electric generators. The drawback to this system is that it requires a lot of land to place the large number of mirrors or other reflectors needed to concentrate solar energy. It takes about one square mile of land to hold enough solar collectors to produce 25,000 kilowatts of electricity. At that rate, about 5,000 square miles of land in the sunny Southwest would be needed to generate enough electricity for the United States today. However, solar proponents point out that agriculture consumes 500,000 square miles of land and produces only about 1 per cent of the nation's energy needs in the form of food.

A number of ideas have been advanced to improve heat efficiency and lessen the need for land. Professor and Mrs. Aden Meinel of the University of Arizona have proposed a solar-thermal system using special lenses to focus sunlight onto chemically coated, nitrogen-filled pipes. The solar farm, as they call it, would occupy 25 square miles and produce 1,000 megawatts of electricity. Another research team, sponsored by the University of Minnesota and Honeywell Inc., has proposed to use a parabolic reflector to concentrate sunlight into air-filled heating pipes. Other researchers are concentrating on what they call "total energy" systems, where solar-thermal plants would make electricity for industrial parks, shopping centers, military bases or even entire towns, while the leftover heat would be used to warm and cool buildings and to run some kinds of equipment.

Wind Power as Indirect Use of Solar Energy

An indirect form of solar energy that can be used to produce electricity is the wind. Wind is created by the earth's "heat engine"—air is heated by the sun's rays and cooled by their absence. The World Meteorological Organization has estimated wind power available at favorable sites around the world at some 20 million megawatts, or more than 40 times the present electric-generating capacity throughout the nation. Wind power has been used for centuries to sail ships and turn windmills. The first windmills appeared in the seventh century A.D. and have been used ever since to pump water, mill grain and—in this century—to produce electricity. In 1941 the world's largest windmill—110 feet high with 175-foot-diameter blades—was built atop Grandpa's Knob near Rutland, Vt. It produced 1.3 megawatts of electricity until it was shut down in 1945.

A joint committee of the National Science Foundation and the National Aeronautics and Space Administration recently suggested that wind power could produce 1.5 trillion kilowatts of electricity by the year 2000 if a development program were actively pursued. That is almost as much electricity as the amount consumed annually in the United States. Today, many experiments to harness wind power are under way in this country and abroad. Engineers are using advanced aeronautical concepts in their search for a slender, lightweight blade that will allow windmill shafts to spin quickly in light winds and be sturdy enough to stand high winds. Last year the federal government began conducting experiments with a 100-kilowatt wind turbine at Sandusky, Ohio. The unit stands 100 feet tall on a steel tower and its propeller blades span 125 feet.

The main problem with windmills for electric-power generation is the expense of construction. At least $5,000 is required to build even a small windmill capable of producing electricity. Consequently, the electricity is likely to be costlier than if obtained from conventional sources. Also, an array of windmills large enough to produce electricity for an entire town or city would take up an enormous amount of space and create a kind of visual pollution of the landscape.

Even so, several wind-power proposals have received considerable attention in recent years. One of the most ambitious is offered by Professor William E. Heronemus of the University of Massachusetts, who suggests that huge windmills be built on floating towers or platforms in the oceans offshore. These would be out of sight from land and could generate large amounts of electricity that could be cabled ashore or, Heronemus suggests, used to separate distilled sea water into hydrogen and oxygen in an electrolytic process. The hydrogen then could be piped or shipped ashore to be burned in electric-generating plants, or someday perhaps used in homes and automobiles in place of natural gas and gasoline.

Progress Toward Tapping Sun Power

D IRECT USE of the sun's energy has been a goal of creative minds through history. Archimedes, the Greek inventor and mathematician, reportedly used the bright metal shields of a thousand soldiers to focus the sun's rays and set fire to an invading Roman fleet during the Second Punic War.[3] A similar tactic was said to have been used during the siege of Constantinople in 626 A.D.[4] During the following centuries, a few experimenters duplicated Archimedes' legendary feat on a smaller scale, setting fire to woodpiles.

In the late 18th century, the French chemist Antoine Lavoisier built a solar furnace that used two large, hollow lenses filled with alcohol to increase their refractive power. The device concentrated the sun's rays so efficiently that Lavoisier was able to melt metals, including steel and platinum, for research purposes. French inventors continued to lead the way in solar experimentation, devising several solar-powered steam engines in the 19th century. One of them operated a printing press that turned out a newspaper appropriately entitled *Le Soleil* (The Sun).[5]

In the United States, the Swedish-American inventor John Ericsson, designer of the Union's ironclad warship *Monitor,* also designed a number of solar-powered engines later in the 19th century. They were the most efficient solar devices built up to that time, but Ericsson concluded that they were still too costly and complex, so he converted them to run on inexpensive coal and gas. A number of other solar-powered engines were built in the early 1900s by various American inventors, but these also proved impractical or uneconomical because of the wide availability of fossil fuels.[6]

One solar device that did find a large market was the solar water heater. In the first half of this century, tens of thousands of these units were manufactured and sold in California, Florida and other sunny states. Most of them employed a rooftop collector consisting of blackened copper tubes in a metal box under a pane of glass, and many had insulated storage tanks to keep water hot overnight. As late as 1950, it was estimated that 50,-000 such heaters were in operation in Miami alone. But by the early 1970s, because of competition from cheap natural gas and fuel oil, the industry had virtually disappeared.

---

[3] According to the Greek historian Galen, this occurred at Syracuse in 212 B.C. However, Livy and Plutarch do not mention the event in their histories.

[4] D. S. Halacy Jr., *The Coming Age of Solar Energy* (1963), p. 197.

[5] Hans Rau, *Solar Energy* (1964), p. 46.

[6] For a detailed historical review, see Wilson Clark, *Energy for Survival* (1974), pp. 361-374.

Solar water pumps and water distillers also found considerable success. From 1870 to about 1910 solar energy provided fresh water for a remote mining area in Chile. Windmills pumped brine into a series of glass-covered troughs where water evaporated and condensed on panes of glass, then trickled down into collection channels. Some 50,000 square feet of solar collectors produced 5,000 gallons of pure water per day in the summertime.[7] An example of a successful water pump was that on the Pasadena Ostrich Farm in California. A large reflector heated water in a steam boiler and powered a pump that raised about 1,400 gallons of water a minute. The installation was famous nationwide in the early 1900s.

Perhaps the best-known American in the field was Dr. Charles Greeley Abbot of the Smithsonian Institution, who has been called the "venerable dean of America's solar scientists."[8] Abbot designed solar cookers, parabolic concentrators, water distillers and steam engines, and was the oldest living U.S. patent holder until his death in 1973 at the age of 101. Another pioneer was Dr. Robert Goddard, who took out several patents on solar-powered devices before be began to concentrate on rocketry in the late 1920s and helped usher in the space age half a century later.

Present Uses of Solar Energy in Other Nations

The use of solar equipment did not die out in many other countries as it did in the United States. In Australia, Japan and Israel, for example, the use of solar water heaters is widespread today. In Australia, the units are used primarily in rural areas, but in Japan and Israel they are found in cities as well. In recent years there have been estimates that more than 200,000 solar water heaters have been installed in Japan and more than 100,000 in Israel.

France leads the world in the large-scale use of concentrated solar energy for scientific research and electricity generation. The French scientist Felix Trombe, a leading solar experimenter, started building solar furnaces in the Pyrenees in the 1950s, and in 1970 completed an eight-story parabolic reflector that heats a furnace to 6,000 degrees Fahrenheit. It has been used primarily to melt metals, ceramics and other materials for research purposes. But in October 1976, the installation was used to power a steam turbine to produce about 100 kilowatts of electricity that were fed into an electric power grid. It was the first time solar energy ever had been used in a commercial utility system.

The French plan to expand the program in an effort to reduce

---

[7] *Ibid.*, p. 364.
[8] By Wilson Clark, p. 367.

the nation's heavy dependence on imported oil and gas. "If there are problems, we'll have to perfect the operation," said Claude Bienvenu of Electricité de France, the national power company. "We have to learn how to use the sun."[9] Also in October, an American corporation, Martin-Marietta, successfully tested a 10-ton, 1,000-kilowatt steam boiler at the French installation in the Pyrenees as part of a program sponsored by the Energy Research and Development Administration (ERDA).[10]

A similar solar-thermal installation began operation in Guanajuato, Mexico, in January 1976, although it does not produce electricity commercially. The facility uses an array of flat collectors to heat water and power a turbine that provides 30 kilowatts of electricity to run two water pumps. These pumps supply a million liters (264,200 gallons) of pure water daily, and the Mexican government plans to build 10 additional solar-powered water-pumping facilities, as well as electricity-generating plants.[11]

Federal Research Effort After Oil Embargo

Despite considerable interest in solar energy among scientists, inventors and entrepreneurs during much of this century, widespread solar development has not occurred in the United States. The main problem has been the high cost of equipment. In addition, there has been continued opposition to solar power from the oil, coal and natural gas industries, which saw their interests threatened. Also, since World War II the nuclear-power lobby has become a strong force in favor of the "peaceful atom" and has opposed substantial commitments to solar energy.

Nonetheless, there have been numerous attempts to encourage American solar development in recent decades. The first federal solar energy bill was introduced in Congress in 1951, dealing with wind power. In 1952, President Truman's Materials Policy Commission issued the famous "Paley Report," named for its chairman, William S. Paley of the Columbia Broadcasting System. This remarkable document strongly recommended solar energy as an alternative to fossil fuels, and predicted that 13 million buildings could be heated by solar units by 1975. It said solar power could meet 10 per cent of the nation's energy needs if the technology were developed aggressively. Needless to say, that optimistic forecast did not come true.

During the next two decades, 10 more bills having to do with solar energy were introduced in Congress, but none passed.

---

[9] Quoted in *Business Week*, Oct. 11, 1976, p. 29.
[10] *The New York Times*, Oct. 18, 1976.
[11] Jeannie Anderson, "Mexico Outstrips U.S. in Preparation of Solar Future," *Critical Mass* (a publication of the Citizens' Movement for Safe and Efficient Energy), May 1976, p. 14.

Then in 1973, the Arab oil embargo suddenly brought solar energy to the forefront of federal energy planning. "The following months were a nightmare of gasoline shortages and a parallel feverish congressional activity toward solar energy legislation," Dan Halacy, an aide to Sen. Paul J. Fannin (R Ariz.) said in a recent report.[12] "Almost overnight solar acquired the sex appeal of a manned space shot and a high promise as the salvation of an energy-starved constituency. To be against solar legislation was practically unthinkable...."

As many as 24 solar-related bills were introduced in the 93rd Congress during 1973 and 1974, and in the fall of 1974 two of the measures became law. The Solar Heating and Cooling Demonstration Act of 1974 established a five-year, $60-million program that directed the National Aeronautics and Space Administration to carry out solar projects and authorized the Department of Housing and Urban Development to supervise the installation of solar units in public and private buildings. The Solar Energy Research, Development and Demonstration Act of 1974 authorized $77-million for a wide variety of programs including a solar resources appraisal, a solar information data bank, demonstration of eight solar projects and establishment of a Solar Energy Research Institute.[13]

In the 94th Congress (1975-76), more than 50 solar measures were introduced, nearly half of which offered incentives for the rapid development of solar equipment by individuals. Although most of the bills did not pass, the clear indication of support for solar development had its effect on budget authorizations. While only $1.4-million in federal funds were spent on solar research and development in 1970, the first budget of the Energy Research and Development Administration contained $40-million for solar projects in 1975. Congressional appropriations topped $100-million in fiscal year 1976 and reached $290-million in fiscal 1977 for various solar programs. "[S]olar proponents find it difficult to believe such good fortune," Halacy wrote. "Indeed, some maintain that they will believe it only when the money actually has been spent...."

In addition, other federal agencies are actively supporting solar projects. The Department of Defense has begun to install solar heating systems at various military bases. The General Services Administration, the government's biggest landlord, is doing the same for certain federal buildings around the nation. The Department of Agriculture has set up a model solar home in South Carolina and is explaining the uses of solar energy through its Extension Service. The Forest Service has used solar

---

[12] "Federal Solar Legislation," a paper presented to the Consumer Conference on Solar Energy Development, Albuquerque, N.M., Oct. 2, 1976.

[13] For details of bills, see *Congressional Quarterly Almanac 1974*, pp. 752-757.

energy at administration buildings, remote maintenance structures and even outhouses. The space agency has installed a solar-powered refrigerator on an Indian reservation in Arizona.

Plan for U.S. Solar Energy Research Institute

One of the most significant federal commitments to solar energy in the long run may be the establishment of a Solar Energy Research Institute (SERI). The Energy Research and Development Administration announced in October that it had postponed the final selection of a site for the new institute until next March. The decision was due to have been announced in December. Competition is fierce among the states for this "research plum."

According to ERDA's official request for proposals, the research institute's mission will be to support the federal solar energy program, help establish an "industrial base" and foster widespread use of solar equipment. During its first year, the institute will have a budget of $4-million to $6-million and a staff of 49 to 76 persons. The figures are expected to grow considerably in subsequent years. About 20 proposals have been submitted by a wide variety of state, university, business and scientific groups. Among the leading contenders are:

Arizona. A state-sponsored Solar Energy Research Commission is working full-time to land SERI; the Battelle Memorial Institute of Columbus, Ohio, would manage and operate it.

California. The state-sanctioned effort is being funded by the Energy Resources, Conservation and Development Commission in connection with several universities.

Colorado. A statewide committee was appointed to prepare Colorado's proposal with the help of legislative appropriations and industry contributions.

Florida. A state energy task force has a full-time staff based at Cape Canaveral.

Michigan. The campaign is led by the Michigan Energy and Resource Research Association, a partnership of state government, universities and industries.

New England. The six New England states (Connecticut, Maine, Massachusetts, New Hampshire, Rhode Island and Vermont) have banded together in an effort supported by the New England Council, an organization of some 2,200 corporations, universities and government officials.

New Mexico. The first state to submit a proposal; it is backed by a group of universities and scientific laboratories in combination with the Stanford Research Institute.

New York. A site near the Brookhaven National Laboratory on Long Island was proposed by a state-university-business group.

Texas. El Paso, San Antonio and Houston all have submitted competing proposals without any state funding or coordination.

Uncertainty in Development Efforts

S OLAR ENERGY's future depends on a complex array of social, political and economic problems whose resolution cannot be predicted with certainty. The most encouraging signs of progress in solar energy today are in industry. More than 250 companies nationwide are involved in solar development, ranging from individual entrepreneurs working in their garages to huge corporations such as PPG Industries, Grumman, Owens-Illinois and subsidiaries of such giant oil companies as Exxon and Mobil. Sales of their products and services probably amount to less than $25-million annually now, but Arthur D. Little Inc., the Cambridge-based consulting firm, estimates that sales of water- and space-heating equipment alone may reach $1.3-billion a year by 1985.[14]

However, even some of solar energy's strongest proponents concede that many predictions of a bright future have been too rosy. "We're still enthusiastic, but we're becoming very realistic," Aden Meinel of the University of Arizona, architect of the "solar farm" concept, told a recent seminar.[15] "Solar sounds so easy to use, but you run into very practical problems. We got pack rats in one of our collectors, for instance, and that's pretty hard to predict on a computer. We've got to keep it in perspective. It needs time and patience."

Similarly, William A. Shurcliff of Harvard University, who annually surveys solar buildings in the United States, has written: "At least 80 per cent of the 200 or so solar-heated buildings that exist today have been uneconomic.... [T]he total lifetime cost of building the equipment and operating it for, say, 20 years is much greater than the money saved through reduction in amount of fuel or electrical power used—assuming that the costs of fuel and electrical power remain at today's levels. In most instances, the overall cost exceeds the overall benefit by a factor of two or three."[16] Nonetheless, Shurcliff believes that a few solar heating systems are "truly economic" and he remains "highly optimistic."

Another solar energy expert who expresses caution is A. I. Mlavsky, head of the Mobil Tyco Laboratories photovoltaic project. "There is a concept—it's engrained in the American ethos—that if you put enough guys and enough money to work on something, you can do anything," Mlavsky said. "It's baloney.... The solar field is a young field.... It's receiving a lot

---

[14] Andrew Tobias, "Solar Energy Now: Why Aren't We Using It More?" *New York*, May 31, 1976, p. 32.

[15] Quoted by Julie Tripp in *The Arizona Daily Star* of Tucson, Oct. 5, 1976.

[16] *Bulletin of the Atomic Scientists*, February 1976, p. 38.

of attention. It will not provide major solutions to our energy problems in a short time, and those—including myself—who sometimes get sucked into thinking that it will are wrong." Mlavsky contends that if the sun is meeting a "significant fraction" of the energy needs in 25 years, "we'll have done a hell of a job."[17]

Barriers of Building Codes and Housing Costs

Among the barriers to the development of solar energy are building codes, financing constraints, tax laws, construction methods and labor requirements. Alan S. Hirshberg, an engineer at California Institute of Technology, wrote recently: "[R]esearch indicates that even when solar space conditioning is economical, institutional barriers to its widespread acceptance will remain."[18] There are about 3,000 building code jurisdictions in the United States, many of them with conflicting regulations. These codes tend to limit the use of new methods and favor existing materials and practices. Few of the codes have yet been revised to list performance specifications for solar equipment. Fire codes also might apply to solar heating systems, requiring fire-resistant parts and adequate insulation around solar units. Health codes might apply to solar systems that used chemicals to transfer or store heat.

Financing limitations of home builders and buyers further inhibit the development of solar energy. The building industry and most home buyers operate on borrowed money, and the high initial construction costs of solar units can make it difficult—and expensive—to get construction or mortgage loans. "The industry is highly sensitive to initial investments (the first cost of its products), and the normal way to reduce the risk of high finance charges is to reduce initial capital requirements," Hirshberg wrote. "Solar devices which have lower operating costs but higher initial investment costs than other energy systems could be expected to meet industry resistance."

The U.S. building industry's general makeup also can be expected to slow solar development to some extent. It is a diverse, fragmented and highly independent industry composed of some 100,000 home builders 90 per cent of whom produce fewer than 100 units each year. The nation's largest builders produce less than 1 per cent of the annual total. So a new technology such as solar energy must be accepted by an enormous number of individual builders and be integrated into existing distribution, sales and service systems. Also, the building industry embraces many crafts and trades, most of them with strong unions. Carpenters, plumbers, roofers, masons and electricians

---

[17] Quoted by Michael Harwood in *The New York Times Magazine*, March 16, 1975, p. 37.

[18] "Public Policy for Solar Heating and Cooling," *Bulletin of the Atomic Scientists*, October 1976, p. 38.

all might have a hand in the installation and servicing of solar systems. Especially complicated would be putting solar equipment into an existing home. This could entail tying into old pipes or excavating a basement to install a heat-storage tank. Even so, some unions see the development of solar energy as a means of increasing jobs.[19]

Another complexity is the question of "sun rights"—what to do if a neighbor's new building blocks the afternoon sun. "Protecting access to sunlight for purposes of solar heating and power generation is a subject awash with unsettled legal questions," two lawyers, Arnold W. Reitze Jr. and Glenn L. Reitze, have written.[20] English law provides a right to daylight, although it is not absolute, while American law is less clear on the subject. Even so, the Reitzes said that proposals to create an absolute right to solar radiation might do more harm than good, and are probably unconstitutional. "To ban shadows is to ban growth," they wrote. "[S]olar energy access legislation is simply not the proper means to attempt to implement no-growth policies, and such laws can be counterproductive to solar use."

Finally, there may be environmental problems associated with large-scale solar development. Efforts to build enormous solar-thermal conversion plants in the Southwest might arouse considerable opposition from local landowners or environmentalists. In the eastern states, the most likely sites for solar-electric plants would be valuable farm, forest or marsh lands. "Solar uses six times more land than nuclear power and that means six times as many battles for sites," R. C. Carlson of the Stanford Research Institute said in a recent paper.[21]

Central Role of the Nation's Electric Utility Industry

Another industry that may play a central role in the development of solar energy is the electric utility industry. This, too, is a diverse industry consisting of about 300 private, investor-owned utility companies that supply 75 per cent of the electricity in the United States and some 3,000 municipal and regional utilities and rural electric cooperatives that provide the remaining 25 per cent.[22] The nation's utilities are far from having a unified position on solar energy, but there probably has been more opposition than support.

On the other hand, some private utilities are investigating the possibility of ownership and control of solar equipment for residential and commercial buildings. The utility would install,

[19] See, for example, "Solar Power Systems," *The Machinist,* November 1976, p. 3. The magazine is published by the International Association of Machinists and Aerospace Workers.

[20] "Protecting a Place in the Sun," *Environment,* June 1976, p. 2.

[21] "Solar Energy: Has the Time Come?" a paper presented to the Consumer Conference on Solar Energy Development, Albuquerque, N.M., Oct. 2, 1976.

[22] See "Future of Utilities," *E.R.R.,* 1975 Vol. I, pp. 185-204.

own and maintain the system, while the homeowner, landlord or business owner would lease it from the utility and make a monthly payment to cover installation and financing costs, maintenance and profit. Public utilities could handle solar energy in much the same way, except the community at large would own the equipment and costs would be lower. Garry DeLoss, an energy specialist who works for Ralph Nader in Washington, D.C., has said: "I'm sympathetic to the idea of publicly owned utilities investing in solar equipment and private utility ownership is a place where the profit motive might be put to good use."

Tax Incentives to Purchase Solar Equipment

To some extent, the future of solar energy in the United States may be determined by tax incentives at federal, state and local levels to promote the use of solar energy. Alan S. Hirshberg has proposed incentives that include federal tax credits, low-interest loans and property tax abatement.

In the past few years, several states have provided property-tax incentives to homeowners to install solar equipment. Property taxes normally go up when home improvements increase assessed value, but the various state measures enable localities to exempt the homeowner from added property taxes on improvements resulting from the installation of solar equipment. According to the National Conference of State Legislatures, states that have passed such measures include Arizona, Colorado, Connecticut, Hawaii, Illinois, Indiana, Kansas, Maryland, Massachusetts, Michigan, Montana, New Hampshire, North Dakota, Oregon and South Dakota.

In addition, state income tax incentives for the use of solar devices are offered by Arizona, California, Hawaii, Idaho, Kansas, Michigan and New Mexico. Georgia, Texas and Vermont exempt solar equipment from state or local sales taxes. Fifteen states provide public funds for solar-energy research, development and promotion through grants to universities, nonprofit groups, public agencies or individuals who undertake specific projects. These states are Arizona, California, Colorado, Florida, Hawaii, Iowa, Maine, Michigan, Montana, Nevada, New Mexico, New York, North Carolina, Ohio and Virginia.

Nevertheless, both state and federal policies at present generally continue to favor fossil fuels and nuclear power over solar energy. In the end the question seems to be: Should the nation continue to deplete fossil fuel resources and expand its commitment to nuclear power, or must it encourage the rapid development of solar energy and other renewable sources? As the greatest energy source on earth, solar power is destined to have its day but no one can say how soon.

Selected Bibliography

Books

Behrman, Daniel, *Solar Energy: The Awakening Science,* Little, Brown & Co., 1976.

Clark, Wilson, *Energy for Survival,* Anchor Books, 1974.

Daniels, George, *Solar Homes and Sun Heating,* Harper & Row, 1976.

Halacy, D. S. Jr., *The Coming Age of Solar Energy,* Harper & Row, 1963.

Keyes, John, *The Solar Conspiracy,* Morgan & Morgan, 1975.

Meinel, Aden B. and Marjorie P., *Applied Solar Energy: An Introduction,* Addison-Wesley, 1976.

Rau, Hans, *Solar Energy,* D. T. Duffin, 1964.

Articles

Abelson, Philip H., "Energy From Biomass," *Science,* March 26, 1976.

Arrandale, Tom, "Solar Energy: Funding Level Debated," *Congressional Quarterly Weekly Report,* April 24, 1976.

Bos, Piet B., "Solar Realities," *EPRI Journal,* February 1976.

Bulletin of the Atomic Scientists, selected issues.

Environment, selected issues.

Faltermayer, Edmund, "Solar Energy Is Here, But It's Not Yet Utopia," *Fortune,* February 1976.

Gilmore, C. P., "Sunpower!" *Saturday Review,* Oct. 30, 1976.

Harwood, Michael, "Energy From Our Star Will Compete With Oil, Natural Gas, Coal and Uranium—But Not Soon," *The New York Times Magazine,* March 16, 1975.

Northcross, Mark, "Who Will Own the Sun?" *The Progressive,* April 1976.

Tobias, Andrew, "Solar Energy Now: Why Aren't We Using It More?" *New York,* May 31, 1976.

Wilhelm, John L., "Solar Energy, the Ultimate Powerhouse," *National Geographic,* March 1976.

Studies and Reports

Carlson, R. C., "Solar Energy: Has the Time Come?" Stanford Research Institute, October 1976.

Eisenhard, Robert M., "A Survey of State Legislation Relating to Solar Energy," National Bureau of Standards, April 1976.

Editorial Research Reports, "New Energy Sources," 1973 Vol. I, p. 185.

Hillhouse, Karin Halvorson, "Solar Energy—Its Environmental Dimensions," Environmental Law Institute, October 1976.

National Conference of State Legislatures, "Turning Toward the Sun, Volume One," January 1976.

Roberson, J. Bob, "The Utility Role in Solar Commercialization," Southern California Edison, October 1976.

POLLUTION CONTROL: COSTS AND BENEFITS

by

John Hamer

**Feb. 27
1 9 7 6**

POLLUTION CONTROL:
COSTS AND BENEFITS

T HE WORDS "ecology" and "economy" come from the same etymological root—from the Greek word meaning "household management." Yet in recent years the two words have taken on conflicting connotations, with environmental protection widely labeled an enemy of economic progress. The drive to stop pollution and clean up the environment, which came to resemble a national crusade in the giddy aftermath of Earth Day 1970, ran head-on into energy shortages, rising inflation, spreading unemployment and deepening recession in the middle years of the decade. Cleaning up the environment and getting the economy back on its feet suddenly were regarded as mutually exclusive. Industry spokesmen, labor leaders and elected officials argued strongly that environmental regulations should be relaxed to stimulate the economy and preserve jobs. Pollution control and environmental improvement were branded as luxuries the nation could ill afford.

Today, however, the tide has turned again, in an unexpected direction. Evidence is mounting that pollution control not only is compatible with economic advancement but actually may contribute to it. Much of the new evidence comes from the federal environmental agencies, which clearly have a stake in promoting pollution control, but some comes from the marketplace. Investments in pollution-control programs and equipment have been found to encourage employment, increase production, create new markets and, on balance, contribute to the national economy.

A booming new pollution-control industry has sprung up to help companies and cities meet environmental standards established by federal, state and local governments. Environmental protection also has been found to benefit the economy in indirect ways—by reducing the health, recreational, agricultural and esthetic costs and damages of pollution. These "external" effects were disregarded for decades as insignificant or unavoidable, but recent calculations have shown their true economic costs to be enormous.

"The lack of adverse economic impact has been the biggest surprise in the unfolding of our programs of environmental reform," John R. Quarles Jr., deputy administrator of the Environmental Protection Agency, said recently. "Industrial extremists typically assumed the worst.... Many industries had predicted widespread plant closings and employee layoffs as a result of the new laws. In fact, the changes have been totally overshadowed by the emergence of environmental expenditures as a positive force in the economy."[1]

One of the most significant recent developments in pollution control was an "Environmental Industry Conference" held in Washington, D.C., in December 1975. Sponsored by the President's Council on Environmental Quality,[2] the conference brought together representatives of more than 200 companies and associations involved in the pollution-control business. In advance of the conference, the council asked two Wall Street analysts to examine the current status and future prospects of the pollution-control industry.

The resulting study, "The Environmental Control Industry—An Analysis of Conditions and Prospects for the Pollution Control Equipment Industry," was prepared by Kenneth Ch'uan-k'ai Leung of F. Eberstadt & Co., Inc., and Jeffrey A. Klein, an independent consultant who has worked for Kidder Peabody and Co., Inc. They reported that several hundred companies were active in pollution control. The study found that industrial, federal, state and local environmental spending, along with associated operating and maintenance expenditures, currently provides more than one million jobs in the United States. Klein and Leung concluded:

> When all factors are considered (including the impact on health and property) there does seem to be an outright economic advantage to pollution control....

A council study in April 1975 on "Environmental Programs and Employment" had similarly found that "the net impact of environmental programs on employment is positive—more people are employed as the result of environmental programs than would be without them. Environmental programs are stimulating construction, equipment and research expenditures that would not otherwise be undertaken."

Efforts to Assess Environmental Cleanup Costs

The council, in its latest annual report, estimated that $21.6-billion was spent on pollution abatement in 1974 *(see table, p. 174)*, averaging $47 per person in the United States. The council further estimated that $381-billion would be spent in the following decade, through the year 1983, doubling the per-person an-

---

[1] Quoted by the Associated Press in *The Washington Star*, Dec. 30, 1975.

[2] With the cooperation and support of the Industrial Gas Cleaning Institute, the National Solid Wastes Management Association, and the Water and Wastewater Equipment Manufacturers Association.

Pollution and the Public

Despite the cost of pollution-control programs, public support for environmental spending has remained strong. The Opinion Research Corporation of Princeton, N.J., a division of McGraw-Hill, in August 1975 released a survey on "Public Attitudes Toward Environmental Tradeoffs." It was based on phone interviews conducted in May and June 1975 with 1,222 persons age 18 and older from across the nation.

The polling organization found that "Even during a time of recession, high unemployment, and rising fuel costs, the public does not voice a readiness to cut back on environmental control programs to solve economic and energy problems." Fully 60 per cent of those surveyed felt it was more important to pay higher prices and taxes to protect the environment than to keep costs down but run the risk of more air and water pollution.

Even so, those surveyed were sharply divided on the question of environment versus employment. They were almost evenly divided—43 per cent agreed and 44 per cent disagreed—on the statement that cleaning up the environment is more important, even if it means closing down some old plants and causing some unemployment.

nual average to $98. In a chapter on "Environmental Economics," the report[3] attempted to assess the environmental costs and economic impact of pollution-control programs. It divided the environmental costs into four categories:

Damage costs—those resulting from direct pollution damages, such as blighted crops, increased illness and higher death rates.

Avoidance costs—the financial and other economic and social costs of attempting to avoid pollution, such as buying air conditioners or moving away from polluted cities.

Abatement costs—the value of resources devoted to reducing pollution, plus indirect effects of such expenditures on economic growth, employment and production.

Transaction costs—the value of resources used in the research, planning, administration, communication and monitoring of pollution control.

The report acknowledged the difficulty of obtaining precise figures, particularly on damage and avoidance costs. "It is disturbing to remain in such a state of ignorance about the various costs and benefits associated with programs to safeguard the environment. Although as a nation we are committed to large expenditures, we do not know whether we are spending too much or too little."

---

[3] "Environmental Quality—The Sixth Annual Report of the Council on Environmental Quality," February 1976, pp. 494-495.

The uncertainty results mainly from ignorance about pollution and its effects on people and the environment. There is uncertainty about how much pollution exists in the nation, what types of pollutants are being emitted and where, how much pollution remains in the environment, how pollutants interact, and how much actual and measurable damage occurs.

"Estimating the dollar cost of painting a pollution-soiled building or of replacing ruined crops is relatively straightforward," the council report said. "But the value of a human life or of a clear sky, or of a place for recreation...cannot be fully translated into dollars.... The fundamental decision that we shall have programs to improve environmental quality is not in question. The crucial question today is not whether to improve the environment, but how much. How much will we gain by increased expenditures, and what are the tradeoffs?"

Assigning a dollar figure to the costs and benefits of pollution control is indeed a tricky exercise. There is no widely accepted method of cost-benefit analysis. Studies done for the Environmental Protection Agency between 1966 and 1975 estimated the annual cost of air-pollution damage in the United States at anywhere from $2-billion to $35.4-billion. Most of the studies said the cost was at least $10-billion a year, however, and two of the most recent said the best estimate was around $20-billion annually. Estimates of annual water-pollution damage also vary widely, but two studies done in 1975 for the agency agreed on the cost figure—that it was about $10-billion a year. According to a draft report of the National Water Quality Commission, made public in September 1975, the cumulative benefits from cleaning up the nation's waters would amount to $12.9-billion by 1980, $36.7-billion by 1985, and $134.2-billion by the year 2000.[4]

The National Wildlife Federation since 1969 has compiled an annual "Environmental Quality Index," attempting to describe numerically the nation's environmental status in seven categories: air, water, wildlife, timber, soil, minerals and living space. According to its latest index[5] estimates:

Annual costs		Annual costs	
If air pollution were controlled	$14.2-billion	If water pollution were controlled	$13.2-billion
Air pollution damages	12.3-billion	Water pollution damages	11.5-billion
Net control cost	$ 1.9-billion	Net control cost	$ 1.7-billion

[4] Klein and Leung, "The Environmental Control Industry," p. 25.
[5] "The 1976 Environmental Quality Index," *National Wildlife*, February-March 1976.

All of the damage cost estimates probably are conservative, since esthetic effects, morbidity, chronic disease and mortality are difficult to evaluate monetarily. Health damage estimates may be especially low since cancer and chronic illness were not considered in most of the studies. Yet evidence is accumulating that a large proportion of cancer cases may be caused by pollutants in the environment or the workplace.

Employment Factor in Pollution-Control Spending

The precise number of jobs created by pollution-control expenditures in the United States is unknown. However, by piecing together various manpower studies by federal, state and industry groups, some general estimates have been made. The Bureau of Labor Statistics recently estimated that for each $1-billion spent on federal pollution-control programs in 1970, some 66,900 jobs were created.[6] Russell W. Peterson, chairman of the Council on Environmental Quality, told the Environmental Industry Conference in December 1975 that $15.7-billion had been spent by government and industry during the year on pollution control. He said that private industry had spent about $10-billion, the federal government around $4.2-billion, and state and local governments the remaining $1.5-billion.

Applying the Bureau of Labor Statistics' rule-of-thumb to Peterson's $15.7-billion figure, it could be surmised that about 1.1 million jobs were created in 1975 by pollution-control spending. However, the council calculates jobs-to-dollars on a different basis—roughly 85,000 jobs per billion.[7] Using that yardstick, the job figure rises to 1.3 billion. Federal laws passed since 1970 to control water and air pollution have been responsible for the creation of most of the jobs. Local governments have been aided particularly by federal funds for the construction of sewers and sewage treatment plants. This work is considered "labor intensive," requiring a relatively large number of workers.

The number of jobs lost in factories forced to close because of pollution-control laws has been far below industry predictions. A recent study by the Environmental Protection Agency's "Economic Dislocation Early Warning System" showed that only 75 plants had closed and 15,710 workers had lost their jobs in the past five years because of environmental regulations—only .016 per cent of the total labor force. Most of the plants

---

[6] The figure 66,900 is an average; the number of jobs varies by the field in which the spending occurs, as follows: 76,700 in research and development, 78,400 in abatement and control operations, 84,000 in radiation control, and 53,600 in waste-water treatment. See "Impact of Federal Pollution Control and Abatement Expenditures on Manpower Requirements," U.S. Department of Labor, Bureau of Labor Statistics, 1975.

[7] The council estimated that up to 25,000 jobs would be created at construction sites, 25,000 in equipment manufacturing and materials processing, and 35,000 from the indirect effect of consumer spending by the new workers.

shut down were old, inefficient and marginally profitable, and probably would have closed even without anti-pollution laws, according to the report. Most of the workers promptly found jobs in nearby plants, it added.

"The problem of plant closings should not be understated, however," the Council on Environmental Quality said in its 1975 annual report. "There is some geographical concentration of the plants which have closed, and many are located in older, industrial towns already suffering relatively high unemployment rates. Their closures can seriously hurt the local economy and people who may have difficulty finding other jobs." Plant closings attributable, or partly attributable, to pollution-abatement costs have been most numerous in these industries: iron and steel (16), food products (11), paper and paper products (10), and chemicals (6).

Rise and Growth of an Environmental Industry

The other—brighter—side of the picture is the new industry that has sprung up to meet the demand for pollution-control equipment. The Klein-Leung study said the industry consisted of at least 600 companies, including more than a dozen "visibly identified" with pollution control, several divisions of larger corporations, and numerous small outfits. Profits have not grown as rapidly in recent years as in the past, Klein and Leung found, largely because of high capital requirements, intense competition, operating inexperience, manpower and materials shortages and inflation. "There are signs, however, that the industry is maturing and that profit margins are stabilizing and are likely to expand in the future," the two analysts reported. "As profits improve, industry participants should be more willing and able to commit resources for capacity expansion."

If total U.S. commitments to pollution-control programs increase, as they seem certain to do over the next decade, the industry can expect sustained growth, most analysts believe. A 1974 study of 12 companies involved in pollution control found that their average sales growth over a seven-year period was 1,775 per cent.[8] Reviewing the industry in 1975, *Barron's* business weekly predicted, "The overall market for gadgetry to clean up the environment is expected to outstrip the growth in the GNP [gross national product] for most of the next decade."[9] *Barron's* cited a study by Frost & Sullivan, a New York market research firm, which projected a 10.5 per cent average annual growth in the market for air-pollution-abatement equipment. The same study said that outlays for waste-water treatment

---

[8] Eileen Kohl Kaufmann, "On Fighting Pollution for Profit," *Business and Society Review,* spring 1974.
[9] David A. Loehwing, "Whiff of Recovery—Pollution Control Has Gone Back Into the Black," *Barron's,* July 14, 1975.

plants by industry and municipalities were expected to climb by about 8 per cent a year.

At the Environmental Industry Conference in December, Richard Love of the Air Pollution Control Group in Stamford, Conn., told the participants: "We've learned that all the environmental markets—air, water, solid waste—are far larger than anyone had imagined, and they also are slower to mature. The greatest opportunities still lie ahead for industry."

Of course, environmental expenditures are perceived in different ways by different industries. Those who benefit from the pollution-control market, such as environmental equipment manufacturers, may ignore the high cost to industries forced to install new controls. On the other hand, those who must comply with strict environmental regulations may consider their investment expenses as money down the drain. Moreover, some industries pollute more than others and thus must spend much more to control pollution. And then there are industries that may both benefit and suffer from strict control requirements. Cement makers, for example, probably will experience increased sales as a result of plant construction for waste-water treatment. At the same time, new cement plants require equipment to control air pollutants.

According to estimates by the Bureau of Economic Analysis in the Department of Commerce, four basic industries— nonferrous metals; pulp and paper; iron and steel; and stone, clay and glass—put more than 10 per cent of their total plant and equipment expenditures into pollution control in 1973-74, the most recent years for which figures were available. This compared to an average of 7.8 per cent for all manufacturing and 4.8 per cent for all businesses. The petroleum and chemical industries also incurred high costs, but the electric utility industry was hit hardest of all by environmental regulations.[10] The nation's utilities spent nearly $3-billion on anti-pollution measures in 1973-74, more than twice as much as the petroleum industry and nearly three times as much as the nonferrous metals industry. And these utility investments are expected to be higher than any other industry's over the next decade.

Varied Reactions of Companies to New Standards

Some companies actually have profited from their experience with controlling their own pollution. Perhaps the most notable example is Dow Chemical Co., which started early and acted aggressively to clean up its wastes. In January 1975, Dow formed a subsidiary, Hydroscience Associates, Inc., to sell its expertise in pollution control. Edwin L. Barnhart, president of the

---

[10] See "Future of Utilities," *E.R.R.*, 1975 Vol. I, pp. 185-204.

new subsidiary, has said its greatest potential growth is in making environmental assessment studies for companies. "I think there's a $100-million market for industrial-pollution control, and no one's exploited more than 10 per cent of it," Barnhart said.[11]

On the other hand, some companies have resisted cleaning up. The U.S. Steel Corp. was called the worst offender by John R. Quarles Jr., in a speech to the Conference Board in New York on Feb. 5, 1976. He said the company's poor record was giving all of industry a bad reputation. U.S. Steel's vice president for environmental affairs, Earl Mallick, said Quarles' statement was "neither factual nor truthful." Mallick said the company had invested nearly $1-billion in pollution control and would "match its environmental progress and programs with other companies and other industries."[12]

Most analyses indicate that the burden of pollution-control requirements on American industry will not be intolerable. Projected investment for environmental purposes by all U.S. industries over the 1974-83 period is unlikely to exceed 6 per cent of total plant and equipment expenditures in any one year and should average about 3 per cent over the decade, according to the Council on Environmental Quality. The council and the Environmental Protection Agency say that demands for pollution-control investments will not disrupt the nation's money markets or displace significant amounts of capital expansion funds. As for inflation, several independent analyses have shown that pollution-control spending increased the rate of inflation by less than 1 per cent a year.[13]

Although there has been some resistance on the part of business and industry to meeting the nation's environmental protection demands, there is evidence that this attitude is changing. Dr. Carl Madden, chief economist for the U.S. Chamber of Commerce, said not long ago: "In the long-range view, maybe it's best for the country if prices go up in industries that are excessive polluters and resource users. Maybe what the ecologists have shown us is that the most economic products are those that result in the least waste." If Madden's attitude spreads throughout the nation's business and industrial community, then America's pollution-control efforts almost certainly will be more successful, and much sooner than expected.

---

[11] "How Dow Sells Its Cleanup Business," *Business Week,* Jan. 20, 1975, p. 74D.

[12] Quoted in *The Wall Street Journal,* Feb. 6, 1976.

[13] The same holds true for future projections. See, for example, "The Macroeconomic Impacts of Federal Pollution Control Programs," Chase Econometric Associates, Inc., January 1975.

Problems of Water-Pollution Control

T HE PRINCIPAL federal law on water pollution is the Federal Water Pollution Control Act Amendments of 1972 (PL 92-500), the most comprehensive and expensive environmental legislation in the nation's history. This law initiated a major change in the basic approach to water-pollution control in the United States by limiting effluent discharges and setting water-quality standards. It set a national goal of eliminating all pollutant discharges into U.S. waters by 1985 and an interim goal of making the waters safe for fish, shellfish, wildlife and people by July 1, 1983. To that end, it required all U.S. industries, by July 1, 1977, to use the "best practicable control technology currently available" for treatment of any discharges. By July 1, 1983, industries will have to use the "best available technology economically achievable."

Limits are based on categories and classes of industries, and are aimed at complete elimination of discharges if technologically and economically possible. Industries may seek relief from the requirements based on alleged lack of economic capability. The law authorized more than $24.6-billion to be spent in cleaning up the nation's waters, including $18-billion in federal construction grants to the states for building waste-water treatment plants. These are allotted on the basis of need, as determined by the Environmental Protection Agency. The federal government pays 75 per cent of the costs and state and local governments provide the remaining 25 per cent. Finally, a pollutant-discharge-permit program was established under strict guidelines administered by the agency.

More than three years after the law's enactment, there is serious question about whether its goals are too ambitious. There are conflicting indications as to how well the requirements are being met by government and industry. On the one hand, the Environmental Protection Agency admits that some 9,000 communities serving 60 per cent of the nation's population will not be able to meet the first-stage deadlines for sewage cleanup. The agency has issued about 26,000 industrial permits for dumping limited amounts of wastes into the nation's waterways, but spot checks showed two out of three permit holders were in violation, according to the National Wildlife Federation. On the other hand, about 5,000 new waste-water treatment plants currently are under construction, and federal grants to cities are rising sharply. In a speech last fall to the Water Pollution Control Federation, EPA Administrator

Estimates of Pollution-Abatement Expenditures

(in billions of dollars)

Pollutant	1974	1983	1974-83 Cumulative
Air Pollution			
Public	0.2	0.8	6.0
Private			
Mobile	5.1	8.8	70.2
Industrial	1.8	8.4	55.6
Utilities	1.0	5.9	34.3
Total Air Pollution	8.1	23.9	166.1
Water Pollution			
Public			
Federal	0.2	0.2	2.3
State and local	7.2	12.6	97.8
Private			
Industrial	2.3	11.2	57.1
Utilities	0.2	1.3	8.1
Total Water Pollution	9.9	25.3	165.3
Solid Waste			
Public	1.5	2.7	21.3
Private	2.1	3.4	28.1
Total Solid Waste	3.6	6.1	49.4
Radiation			
Nuclear Power Plants	—	0.05	0.2
Grand Total	21.6	55.3	381.0

No estimates available for strip-mined land reclamation and noise control.

SOURCE: Council on Environmental Quality

Russell E. Train said: "Over 97 per cent of all water dischargers are either now in compliance with pollution-control standards or are on definite water cleanup schedules...."[14]

One of the biggest problems with the program involved the $18-billion authorized for construction of sewage treatment plants. In 1972, the Nixon administration impounded most of this money, and only when the Supreme Court early in 1975 ruled the impoundment illegal was a substantial amount released to states and localities. These grants fell from $1.6-billion in 1973 to $1.4-billion in 1974, then jumped to $3.6-billion in 1975 and are expected to climb to $5.2-billion in 1976 and $6.2-billion in 1977.

[14] Speech in Miami Beach, Fla., Oct. 8, 1975.

Another provision of the 1972 act established a 15-member National Commission on Water Quality to investigate the problems of achieving the 1983 goals and to report to Congress and the nation. That group's final report is due to be published later in 1976. A draft report made public in the fall of 1975 said that cleaning up the nation's lakes and rivers would cost industry and government between $97-billion and $130-billion by 1983, considerably lower than the Council on Environmental Quality's estimate of $165-billion.

The draft report said that such a cleanup would result in higher prices for many consumer items and would cause some unemployment and some factory shutdowns. But it said the benefits of such expenditures would include substantial improvements in commercial and recreational fishing, wider availability of beaches and swimming areas, and increased property values for land near water. The economic benefits from opening more beaches and improving fishing would total more than $27-billion by 1985, the report said. The draft report did not mention health aspects of water pollution. If health factors were included, the potential benefits of water cleanup expenditures would be much higher. A staff estimate put the cost of water cleanup at $44 per person a year for the next 10 years.

Shortly after the draft report was made public, a White House task force sharply criticized the work of the National Water Quality Commission in another report. The task force, a part of the Domestic Council, said the commission had underestimated costs and overestimated benefits in its draft. However, several commission staff members said the White House task force was biased in favor of industry, pointing out that 11 of its 26 members were from the Department of Commerce while only 5 were from environmental agencies.[15]

Pollution From Farms, Mines and Construction

Even so, there have been other suggestions that some of the costs of water cleanup are being underestimated. In an analysis in *Power Engineering* magazine, F. C. Olds said government figures showing total cleanup costs over the next decade to be more than $300-billion do not include "costs to cope with acid mine drainage, agricultural feed lot runoff, urban storm water runoff problems, noise levels, the ultimate zero discharge, and numerous other matters relating to environmental protection." He called the present laws "unworkable" and "unrealistic."[16]

Some government environmental officials also see enforcement difficulty ahead. Gary Dietrich of the Office of Water and

---

[15] *The Washington Star.* Oct. 10, 1975.

[16] "Environmental Cleanup 1975-1985: Huge New Costs, Little Benefit," *Power Engineering.* September 1975, p. 38.

Hazardous Materials has said: "We are in the finishing stages of the first round of our battle against water pollution. The second round will be far more difficult because we will be dealing with toxic pollutants which we know little about."[17] The Environmental Protection Agency has found excessive levels of toxic heavy metals such as mercury, cadmium, manganese, lead and iron in most major American river systems. They have concluded that "non-point" sources—rainwater, storm sewers, and agricultural runoff—are responsible for much of this contamination. The Council on Environmental Quality has estimated that it would cost $235-billion to control storm-sewer runoff alone—nearly half again as much as the projected expenditures for all other types of water-pollution control during the 1974-83 decade.

Train, in his speech to the Water Pollution Control Federation, pointed out that more than 400 million acres in the nation are crop land, from which two billion tons of sediment flow annually into lakes and streams. This includes much of the 440 million pounds of toxic pesticides used every year, as well as nitrogen and phosphorus from the 41 million tons of fertilizers used every year. In addition, enormous quantities of animal wastes enter the nation's waters. Livestock produce 10 times more waste than do humans. Five to 10 per cent of the total sediment load in the nation's rivers comes from the 10 to 12 million acres of commercial forest harvested every year.[18]

Strip mining is another source of pollutants, affecting some 350,000 acres annually. In northern Appalachia alone, mine drainage discharges more than one million metric tons of acid into surface and ground waters every day.[19] Construction and excavation related to urban sprawl, which consumes hundreds of square miles per year, generate even more sediment than agricultural activities. Congress placed primary responsibility for the management of "non-point" source pollution with the states, and there is considerable variation in the efficacy of state programs.

Mixed Results From Air-Pollution Fight

THE ORIGINAL deadline for cleaning up the nation's air, under the Clean Air Act of 1970, was May 31, 1975. By that date, the air was supposed to be "safe enough to protect the public's health." Just before the deadline, Train confessed: "Despite significant progress, a number of the nation's 247 air-

---

[17] "A Turn in the Tide—Pollution Battle Being Won?" *U.S. News & World Report,* Aug. 4, 1975, p. 57.

[18] See "Forest Policy," *E.R.R.,* 1975 Vol. II, pp. 865-884.

[19] See "Strip Mining," *E.R.R.,* 1973 Vol. II, pp. 861-881.

quality-control regions will not meet all of the air quality standards." In two out of every three regions, pollution levels were higher than those specified by the 1970 act.

However, the air in most U.S. cities today is considerably cleaner than it was five years ago, even if it does not meet the strict standards of the act. Concentrations of sulfur dioxide have dropped 25 per cent nationwide since 1970, including a 50 per cent decrease in major metropolitan areas, according to the Environmental Protection Agency. Particulate matter—dust, smoke and soot—has decreased by 14 per cent nationally over the same period. Automobile exhaust pollution from 1975 model cars was as much as 80 per cent below that from 1967 cars of comparable weight and engine size. Still, auto emissions remain one of the largest sources of urban air pollution today.

Why the mixed results in fighting air pollution? "No one back in December of 1970 imagined that it would be easy to achieve clean air," Train has said. "However, many of us doubtless underestimated the complexities involved, and certainly few foresaw...the worldwide energy crisis and economic recession."[20] States are hampered by lack of funds to hire inspectors to enforce the laws. Even so, both the Environmental Protection Agency and the Council on Environmental Quality are continuing to push the message that the costs of cleaning up the nation's air are well worth the benefits that will be derived.

The National Academy of Sciences has estimated that air pollution causes 15,000 deaths and seven million sick days a year. The EPA has estimated that medical costs, plus lost working days, could total as much as $7.6-billion a year. In addition, air pollution damages property, clothing and other materials. The agency said pollution causes urban families to pay up to $57 a year to clean and replace soiled clothes and $20 a year to repaint houses and automobiles. In the past, air pollution was not a significant problem in rural areas, but that may be changing. In 1975, the agency said that dangerous levels of smog had been detected in broad regions of the eastern United States. In some instances, pollution levels were worse in communities 50 miles beyond major cities than in the cities themselves. Sometimes this is attributable to heavy commuting traffic and sometimes to electric power plant smokestacks.

Deadlines for Auto Emissions, Stationary Sources

Deadlines for the final standards on three automobile exhaust pollutants—carbon monoxide, hydrocarbons and nitrogen oxides—have been delayed three times and are now set to take effect for 1978 model cars. Most American manufacturers have

---

[20] Quoted in *U.S. News & World Report,* Aug. 4, 1975, p. 58.

turned to the catalytic converter to clean up exhausts, but under some conditions these devices have produced sulfuric acid mist—another potentially dangerous pollutant.[21] President Ford has asked Congress to delay the imposition of the original emission standards and to keep the present levels in effect through the 1981 model year. In the Senate, a bill was reported by the Public Works Committee early in February 1976 to grant a one-year delay, until 1979, for automakers to meet the carbon monoxide and hydrocarbon standards.

As for stationary sources of air pollution, such as industrial smokestacks and electric power plants, the big fight has been over the installation of "scrubbers" to remove hazardous pollutants from emissions. The Environmental Protection Agency has said that of 220,000 stationary sources of noxious fumes in this country, 20,000 account for 85 per cent of the emissions. The agency is concentrating on these primary polluters and has investigated most of them, achieving compliance from more than 80 per cent. However, some 3,000 major polluters still are in violation of the law, and a concerted effort to clean up their pollution is now under way. Electric utilities have been pushing for adoption of emission standards that could be met by the construction of tall smokestacks (to disperse pollutants over a wide area), and by use of so-called "intermittent" standards, which entail closing down plants only when an atmospheric inversion or other weather condition creates a pollution hazard.

The EPA, however, is insisting on the installation of scrubbers, which it claims have been operating reliably for a year or more at several places around the nation. About 100 power plants already are committed to scrubbers despite the high cost of installation but 150 to 200 others have resisted scrubber technology. The Industrial Gas Cleaning Institute, which represents scrubber manufacturers, estimates that scrubbers cost about half as much as the measurable damages caused every year by the fumes they would eliminate. The Senate Public Works Committee bill would require utilities and factories to meet compliance schedules, although those under order to burn coal because of oil shortages could receive extensions.

Congressional Proposal to Aid Urban Compliance

As for general air-quality standards in the nation's cities, the initial goal of the 1970 act has been found unrealistic. At least 30 major cities still would have polluted air even if they met all automobile, industry and utility standards. These cities would have to take more drastic measures—such as restricting traffic

---

[21] See "Auto Emission Controls," *E.R.R.*, 1973 Vol. I, pp. 289-312.

or shortening work weeks—to bring their air-pollution levels into compliance. But even though many Americans express general support for limiting cars in downtown areas, improving mass transit, encouraging car pools and establishing exclusive bus lanes, most individuals remain resistant to changing their driving habits. The proposed Senate bill would allow cities up to 10 extra years to meet the 1977 compliance deadline, provided they adopted "reasonable transportation control measures" such as bus lanes and car inspection programs.

Another major issue in the air-pollution fight is that of "nondegradation"—the concept that air quality should not be allowed to deteriorate in sparsely populated areas where it is already cleaner than the law requires. As more and more regions are viewed as desirable for development, this will be a growing controversy. Environmentalists argue that the air should be kept as pure as it now is in such areas, while developers contend that the air should be permitted to reach a pollution level comparable to that in other populated areas. The Senate bill would allow states to set up a classification system for non-polluted areas, with strict standards for national parks, wilderness areas and other pristine regions, and lesser standards elsewhere.

The bill also would set up a procedure to protect workers against "environmental blackmail"—getting laid off or fired by employers who blame the cost of meeting clean air standards—by allowing them to demand a hearing before the Secretary of Labor. The EPA also would be able to hold public hearings on plant closings that are attributed to the cost of compliance with federal air-pollution-control regulations. The House is far behind the Senate in its consideration of Clean Air Act amendments, but it is expected to deal with many of the same issues in its version of the legislation.[22]

Questions of Solid Waste and Land Use

OFTEN IGNORED in analyses of the costs and benefits of pollution control are the problems of solid waste and land use, both of which are enormously complicated issues and are intertwined with air and water pollution. More than 135 million tons of solid wastes are generated by America's households, stores and office buildings annually—about 1,000 pounds a year for every man, woman and child—and the amount is growing by 3 to 4 per cent a year. If agricultural, industrial, construction,

---

[22] See *Congressional Quarterly Weekly Report*, Feb. 14, 1976, p. 311.

Capital Investments for Pollution Control
(in billions)

	1968	1970	1972	1974
Total Capital Investments	$67.77	$79.71	$88.44	$112.40
Pollution-Control Investments	1.13	2.50	4.50	6.92
Per Cent of Total	1.7%	3.1%	5.1%	6.2%

SOURCE: McGraw-Hill Publications Company

sewage and mining wastes are included, the total exceeds 4.5 billion tons a year. The amount of solid waste discarded per person in the United States has doubled in the past 50 years, and is growing about five times faster than the population.[23]

This solid waste is such a problem because traditional means of refuse disposal cannot keep up with the burgeoning amounts of garbage and trash. About half of the nation's cities are running out of available land for waste disposal, according to a 1974 survey by the National League of Cities. Open city dumps are being closed by the Environmental Protection Agency because they create air pollution when trash is burned, water pollution when it rains, and other health and esthetic damage. Sanitary landfill is an alternative used by about 80 per cent of the nation's cities, but relatively few operate within accepted standards, which require that garbage and trash be compacted and covered with a layer of earth by bulldozers or dump trucks each day. No burning is allowed.

Development of Resource Recovery Technology

Large-scale resource recovery is being touted by many leaders of industry, science and government as the best answer to the solid-waste crisis. Garbage is collected by trucks in the normal manner, then ground up in a shredder and run through a magnetic separator which removes metals that can be recycled by steel mills. Glass and other materials also may be removed for recycling. The remaining shredded trash is mixed with coal or other fuels and can be burned to produce steam in electric power plants.

About 50 cities today are in some stage of commitment to resource or energy recovery, although only a few systems actually are operating or under construction. Some have been criticized for high air-pollution levels. The new technology shows promise, but the cost of building recovery facilities is high, and most of the plants built so far have relied on subsidies.

[23] See "Solid Waste Technology," *E.R.R.*, 1974 Vol. II, pp. 641-660.

Making them self-supporting is difficult because the market for scrap materials is cyclical. Some analysts believe that "source reduction"—cutting down on excess packaging and eliminating throwaway containers—is preferable to building elaborate resource-recovery plants and would be less costly.

Another difficult land-pollution problem is sludge, the mucky residue of sewage treatment. Until recently, sludge usually was burned or buried, or dumped into bays or oceans. But today, because of population expansion and improved sewage treatment, there is much more sludge around. And stiffer environmental standards are making it harder to dispose of. "Sludge is the most serious dilemma we face in waste-water treatment," an EPA official has said.[24] "It's Catch-22. The cleaner we make the water, the more sludge we create." Sludge often contains toxic materials such as heavy metals and pesticides, as well. Nationally, about four million tons of sludge are generated annually, and that total is expected to reach 10 million tons by 1985. Cities are experimenting with a variety of disposal methods, such as converting sludge to methane gas, burning it under pressure, and dumping it on crop lands as fertilizer. But there are problems with all of these methods, and the search continues for an environmentally and economically sound solution.

Issues of Land Use Planning and Urban Sprawl

Air pollution, water pollution, solid waste and sludge pollution are all related closely to the way the United States uses its land. Land use is the realm in which all other forms of pollution come together, for without careful land-use planning, effective pollution control is difficult if not impossible to achieve. Urban sprawl, for example, almost always generates air, water and solid waste pollution, while unplanned development in rural areas can do the same thing. Congress, after repeated attempts, has not passed a comprehensive national land-use planning bill, although 17 states enacted land-use laws in 1975.

The land-use issue exemplifies the entire range of arguments over pollution control. The perception of costs and benefits varies among different people. While some believe that environmental protection must take precedence, others insist that economic considerations should hold sway. Polarization between the two extremes is quick to develop, with the result that the controversy breaks down into a hopeless battle of true believers. What the new analyses of environmental economics are indicating, however, is that there is no need for the nation to choose between a clean environment *or* a healthy economy. It should be possible for the United States to have both.

---

[24] Unidentified official quoted in *The Wall Street Journal,* Dec. 16, 1975.

Selected Bibliography

Books

Auld, D. A. L. (ed.), *Economic Thinking and Pollution Problems*, University of Toronto Press, 1972.

Baxter, William F., *People or Penguins—The Case for Optimal Pollution*, Columbia University Press, 1974.

Goldman, Marshall I. (ed.), *Controlling Pollution—The Economics of a Cleaner America*, Prentice-Hall, 1967.

Kneese, Allen V. and Charles L. Schultze, *Pollution, Prices, and Public Policy*, Brookings Institution, 1975.

Kneese, Allen V., Robert U. Ayres, and Ralph C. d'Arge, *Economics and the Environment: A Materials Balance Approach*, Johns Hopkins University Press, 1970.

Articles

"EPA Aims to Preserve Profits While Protecting Environment," *Commerce Today*, Nov. 24, 1975.

Heffernan, Patrick, "Jobs and the Environment," *Sierra Club Bulletin*, April 1975.

Loehwing, David A., "Whiff of Recovery—Pollution Control Has Gone Back Into the Black," *Barron's*, July 14, 1975.

McWethy, Jack, "Now, Second Thoughts About Cleaning Up the Environment," *U.S. News & World Report*, Jan. 19, 1976.

National Wildlife, selected issues containing annual Environmental Quality Index.

Olds, F. C., "Environmental Cleanup 1975-1985: Huge New Costs, Little Benefit," *Power Engineering*, September 1975.

"The Surprisingly High Cost of a Safer Environment," *Business Week*, Sept. 14, 1974.

Studies and Reports

Abel, Fred H., Dennis P. Tihansky and Richard G. Walsh, "National Benefits of Water Pollution Control," Environmental Protection Agency, 1975.

Biniek, Joseph P., "The Status of Environmental Economics," report by the Environmental Policy Division, Congressional Research Service, Library of Congress, for the Senate Public Works Committee, June 1975.

Council on Environmental Quality, "Environmental Quality—1975," sixth annual report and selected previous reports.

—"The Economic Impact of Environmental Programs," November 1974.

—"The Macroeconomic Impacts of Federal Pollution Control Programs," January 1975.

Environmental Protection Agency, "Evaluation of Techniques for Cost Benefit Analysis of Water Pollution Control Programs and Policies," December 1974.

Leung, Kenneth Ch'uan-k'ai and Jeffrey A. Klein, "The Environmental Control Industry—An Analysis of Conditions and Prospects for the Pollution Control Equipment Industry," December 1975.

Opinion Research Corporation, "Public Attitudes Toward Environmental Tradeoffs," August 1975.

Real Estate Research Corporation, "The Costs of Sprawl—Detailed Cost Analysis," April 1974.

Waddell, T. E., "The Economic Damages of Air Pollution," Environmental Protection Agency, 1974.

INDEX

Significance of public participation in science - 120

Cocaine. See Drugs, Drug Addiction

Committee for the Scientific Investigation of Claims of the Paranormal - 48

Computer Crime
Access to stored and transmitted data - 70
Analyses of 150 cases - 73, 74
Case of the missing boxcars - 78
Computer theft and blackmail - 68, 70
Congressional action - 64
Difficulty of detection and conviction - 64
Efforts to improve safeguards, training - 72
Electronic theft, fraud and embezzlement - 66
Equity Funding scandal - 68, 69, 71, 72
Leniency for white-collar offenders - 75
Predictions of rise in computer crime - 78
Reported cases of computer misuse (box) - 67
'Round down' fraud - 70
Sabotage and vandalism by employees - 65
Virtual absence of computer security - 71

Copernicus, Nicholas - 10

Council on Environmental Quality (CEQ)
Air pollution control - 177
Environmental programs and employment - 166, 169, 170, 172
Pollution-abatement expenditures (1974-1983) - 174
Water pollution control - 175, 176

D

DeWied, David - 27, 28
Dianetics - 57
Disaster Relief Act of 1974 - 129, 140
Dixon, Jeane - 54, 55
DNA research. See Genetic Research
Drugs, Drug Addiction
Alcohol and tobacco - 90
Anti-cancer drugs - 95, 96, 97
Anti-psychotic drugs - 33
Cocaine - 31
Developing theories of brain chemistry - 26, 27, 39
Endorphin testing with mental patients - 28, 29
Ethics of psychotropic drugs use - 38
Heroin - 26
Laetrile and Krebiozen - 96, 97
Mood-altering agents - 23, 38
Natural painkillers - 24-27
Strychnine - 27, 28
Valium and Librium - 38

E

Earthquake Forecasting
Advances in Forecasting Research
American program funding - 131
Chinese achievement - 125, 130
Concern over public reaction to warnings - 128, 137, 138
Development of techniques - 129
Early warning signals - 136, 137
California Forecast
Plea for more funds for American program - 131
Prediction of impending earthquake - 126
Public response to forecast - 138
Possibility of earthquake control - 141
San Andreas Fault (map) - 127
Earthquake Causes and Consequences
Devastating effects in the world for centuries - 132
Dilatancy theory - 137
Early attempts at forecasting - 134
Earthquakes in 1976 - 133
Memorable earthquakes in the United States - 133
Plate tectonics theory - 139
Earthquake-risk Zones
Impending California earthquake - 126
In the United States (map) - 135
Methods used to identify zones - 129, 130
San Andreas Fault - 127
Hazard-reduction Programs
Earthquake control or modification goal - 140
Low-interest loans and relief money - 140
National Academy's recommendations - 139

Economic Forecasts. See also Politics of Science
Computer crime costs - 63
Environmental cleanup costs - 166
Jobs factor in pollution-control spending - 169
Long-term decline in research funding - 6
Patent right disputes - 13, 111
Solar energy's future - 158
Uncertainties about earthquake warnings - 137, 138

Einstein, Albert - 16

Electric Utilities
Air pollution control - 178
Solar energy development - 160

Electronic fund transfer (EFT) system - 76

Energy Research and Development Administration (ERDA)
Solar energy programs - 146, 155-157

184

T, U, V

Toxic Substances
Cancer and the environment - 90, 93
Pesticides danger - 17, 94, 176
Sewage treatment residue - 181
Toxic Substances Control Act of 1976 - 111
Water pollution from farms, mines and construction - 175, 176

Unidentified Flying Objects (UFOs)
Jimmy Carter's 1973 UFO report - 45
Occurrences often mistaken for UFOs - 46
Renewed argument over UFO sightings - 44

U.S. Chamber of Commerce
Computer crime report - 63

U.S. Geological Survey
Concern over reaction to earthquake forecasts - 128, 129

Earthquake control or modification - 140
Earthquake predictions - 125, 126
Earthquakes in 1976 - 133

W-Z

Water pollution. See Pollution Control
Wehr, Thomas - 34
Werewolves, warlocks, and witches - 51, 52
Whitcomb, James H. - 126, 127, 136, 139
Whiteside, Thomas - 64, 71, 72, 77, 79
Wind power - 151
World Meteorological Organization - 151